AFTER BORDER EXTERNALIZATION

AFTER BORDER EXTERNALIZATION

Migration, Race, and Labour in Mauritania

Hassan Ould Moctar

BLOOMSBURY ACADEMIC
LONDON • NEW YORK • OXFORD • NEW DELHI • SYDNEY

BLOOMSBURY ACADEMIC
Bloomsbury Publishing Plc
50 Bedford Square, London, WC1B 3DP, UK
1385 Broadway, New York, NY 10018, USA
29 Earlsfort Terrace, Dublin 2, Ireland

BLOOMSBURY, BLOOMSBURY ACADEMIC and the Diana logo
are trademarks of Bloomsbury Publishing Plc

First published in Great Britain 2024

Series design by Grace Ridge
Cover images © Associated Press via Alamy Stock Photo
and Joanna McCarthy via Getty Images

A catalogue record for this book is available from the British Library.

Library of Congress Cataloging-in-Publication Data
Names: Ould Moctar, Hassan, author.
Title: After border externalization : migration, race, and labour in Mauritania /
Hassan Ould Moctar.
Description: 1. | New York : Bloomsbury Academic, 2024. |
Includes bibliographical references and index.
Identifiers: LCCN 2024008631 (print) | LCCN 2024008632 (ebook) |
ISBN 9781350376786 (hardback) | ISBN 9781350376823 (paperback) |
ISBN 9781350376793 (epub) | ISBN 9781350376809 (ebook)
Subjects: LCSH: European Union–Mauritania. | Foreign workers–Mauritania. | Ethnology–
Mauritania. | Mauritania–Emigration and immigration. |
Mauritania–Politics and government. | Mauritania–Boundaries.
Classification: LCC HD8830.M63 M638 2024 (print) | LCC HD8830.M63 (ebook) |
DDC 331.6/2661–dc23/eng/20240416
LC record available at https://lccn.loc.gov/2024008631
LC ebook record available at https://lccn.loc.gov/2024008632

ISBN: HB: 978-1-3503-7678-6
ePDF: 978-1-3503-7680-9
eBook: 978-1-3503-7679-3

Typeset by Integra Software Services Pvt. Ltd.

To find out more about our authors and books visit www.bloomsbury.com
and sign up for our newsletters.

BLOOMSBURY
OPEN ACCESS

An ebook edition of this book is available open access on bloomsburycollections.com. Open access was funded by the Bloomsbury Open Collections Library Collective.

Bloomsbury Open Collections is a collective-action approach to funding open access books that allows select authors to publish their books open access at no cost to them. Through this model, we make open access publication available to a wider range of authors by spreading the cost across multiple organizations, while providing additional benefits to participating libraries. The aim is to engage a more diverse set of authors and bring their work to a wider global audience.

More details, including how to participate and a list of contributing libraries, are available from http://www.bloomsbury.com/bloomsbury-open-collections.

To my family, for helping me look beyond seeking definitions in difference.

CONTENTS

FIGURES AND TABLES

Figures

Tables

ACKNOWLEDGEMENTS

All praise and thanks are foremost due to Allah, the Divine Essence from which all else flows and to which all ultimately returns.

In the preface to *How Europe Underdeveloped Africa*, Walter Rodney forgoes the ritual avowal that 'all mistakes and shortcomings are entirely my responsibility'. Dismissing such claims as 'sheer bourgeois subjectivism', he instead emphasizes that 'responsibility in matters of these sorts is always collective, especially with regard to the remedying of shortcomings'. Strengths and shortcomings alike are indeed the fruit of collective effort. In the case of the colleagues and friends mentioned below, this collective responsibility is no less applicable to those pieces of advice and criticism which, against my better judgement, I didn't follow, since they forced me to be crystal clear in my own mind about approaches taken and choices made. That said, they probably shouldn't stand accused of responsibility for some of the more glaring shortcomings.

The PhD thesis upon which this book is based would barely have gotten off the ground, let alone made it to this stage, were it not for the consistent intellectual guidance of my former PhD supervisor and postdoc mentor Paolo Novak. His mentorship has been proactive, critical, and systematic, though without ever dictating the course of the research process. I will forever fondly recall our many conversations in his office, all of which were marked by his characteristic blend of informal warmth and searing judicious critique. The intellectual support and words of encouragement of my second and third PhD supervisors at SOAS, Adam Hanieh and Anna Lindley, have also been invaluable. I am also grateful to the National University of Ireland for funding the PhD research upon which this book is based, and to the Economic and Social Research Council for the postdoctoral fellowship that allowed me a year after my PhD to gather some thoughts in the form of publications.

I am grateful to many in Mauritania for their generous assistance during my PhD fieldwork, particularly Ahmed Haroun Cheikh Sidiya, Haimoud Ramdane, Zeidane Mohamed Abdellahi, Abdellahi Aweh, and Sidi Mohamed Ould Beidy. I am also indebted to Badiane Ibou, Ousmane Touré, Maître Fatimata M'Baye, ElHadj Mbow, and all of the wonderful people at the AMDH, for being unwaveringly accommodating and supportive of my research during fieldwork. My field visits beyond Nouakchott would not have amounted to much were it not for the kind help of Daouda Saar of the AMDH in Rosso and Mohammed Ould Ely in Nouadhibou. And these trips would not have happened at all without the generosity and hospitality of Ismail M'Bareck and his family in Rosso, and Ahmed Ould Neini in Nouadhibou.

I am forever grateful for my uncles, aunts, and cousins in the Alweimine family in Mauritania for the care, support, and life lessons they offered me long before this project's formal commencement. In this regard I am particularly grateful to my uncle Abdellahi and his family – Aicha, Moctar, Ahmed, and Fadilla – and to my uncles Yacoub and Hassan. I am also thankful to my uncles Mohamed and Sid 'Ahmed and their families, as well as to Adiya, Ismail, and Naha, and all of the *lkhiyâm* in Boutilimit and Ain Salama.

This book has also been helped along by a variety of exchanges and forms of support spanning workshops, conferences, and institutions I've found myself in over the years. I am thankful to Sebastian Chauvin and Manolis Pratsinakis, my MSc supervisors at the University of Amsterdam, and I am deeply indebted to John O'Brennan at Maynooth University in Ireland. My PhD examiners Ruben Andersson and Hannah Cross vastly improved the intellectual breadth and tone of this work in its transition from a PhD thesis to a book. I am also thankful to friends in my SOAS Development Studies PhD cohort, and in particular to Karen Iversen, Iris Lim, and Shona Macleod for the helpful feedback and criticism over the years, and to Enrique Ballivian, Kira Brenner, and Joe Buckley for their incisive Marxist (or is it Marxian?) insights. Thanks also to Ben Reynolds, Oli Harrison, and Oisín Fagan for grounding comments and suggestions. The West Africa working group of the migration-control.info network has also been a consistent source of inspiration, in particular fellow coordination team members Moctar Dan Yaye, Leonie Jegen, Aino Korvensyrjä, and Laura Lambert.

While by no means exhaustive, the following people have also nourished this book through either reading portions, offering advice, or the exchange of ideas: Nora El Qadim, Chouki El Hamel, Lorena Gazzotti, Philippe Frowd, Joe Turner, Ċetta Mainwaring, Maurice Stierl, Martina Tazzioli, Nandita Sharma, Manuela Boatcă, Aparna Kolar, Daria Krivnos, Elisa Pascucci, Sebastian Cobbarubias, Maribel Casas Cortes, Martin Lemberg Pederson, Beste İşleyen, Kerem Nişancıoğlu, Gilbert Achar, Matthew Porges, July Blalack, David Keen, Rajesh Venugopal, and Philip Donnery. Thanks are also due to Nick Wolterman, Nadine Staes-Polet, Kumaraguru Elangovan, and Olivia Dellow at Bloosmbury, for their consistent guidance and patient responses to my panicked last-minute queries.

I owe more than I will ever know to a small circle of friends in Ireland, who have kept me centred with the tunes and the chats, the craic and the mischief. I am also grateful to many within what can loosely be called the Irish left – in particular the anti-racist movement, the housing movement, and the Ireland Palestine Solidarity Campaign – whose camaraderie, solidarity, and political education over the years have been deeply formative. Infinite gratitude to my parents Moctar and Detta for patiently bearing with me as I moved from place to place trying to figure out how to best get this thing done, and for all of their love, support, and encouragement in the process. I am also eternally grateful to Ayesha, who is entirely responsible for my exercising the discipline necessary to see this work to completion, and who had a hand in one or two pieces of its intellectual content as well.

Finally, I am in equal measure indebted to and inspired by the many migrant workers and Mauritanian nationals who shared their time, experiences, and delightful company with me over the course of my PhD fieldwork. I will be forever in awe of the strength, wit, and grace with which they and countless others navigate merciless conditions of violence and abandonment.

NOTE ON TRANSLITERATION AND TRANSLATION

Words and phrases in the *ḥassânîya* dialect of Arabic have been transliterated using Brill's simple Arabic transliteration system. These *ḥassânîya* words and phrases all appear in italics, with the exception of the names of public figures. Where precolonial public figures are concerned, this same transliteration system is used. Where more recent public figures are discussed, I spell their names using the prominent French rendering. French spelling is also used for place names in Mauritania. I have translated all quotes from French oral and written sources into English.

ABBREVIATIONS

ACP-EU	Africa Caribbean Pacific–European Union Partnership Agreement
AFCF	Association des femmes chefs de famille
AFD	Agence Française de Développement
AMDH	Association mauritanienne des droits de l'homme
ANRPTS	Agence nationale du registre des populations et titres sécurisés
AOF	Afrique Occidentale Française
APEAH	Association pour l'environnement et l'action humanitaire
AU	African Union
CCMM	Cadre de concertation des migrants en Mauritanie
CFA	West African CFA franc (currency)
CMJD	Conseil militaire pour la justice et la démocratie
DGSN	Direction Générale de Sûreté Nationale
ECOWAS	Economic Community of West African States
EDF	European Development Fund
EEAS	European External Action Service
ENP	European Neighbourhood Policy
EU	European Union
EUROSOR	European Border Surveillance System
FAMAM	Fédération associations des migrants de l'Afrique de l'Ouest en Mauritanie
FAO	Food and Agriculture Organization
FDI	Foreign Direct Investment
FIIAP	International Ibero-American Foundation for Administration and Public Policies
FLAM	Forces de libération des africains de Mauritanie
FONADH	Forum national des ONG de droits de l'homme
FPA	Fisheries Partnership Agreement
Frontex	Frontières extérieures
GAM	Global Approach to Migration
GAR-SI Sahel	Groupes d'action rapide – surveillance et intervention au Sahel
GIZ	Deutsche Gesellschaft für Internationale Zusammenarbeit
ICMPD	International Centre for Migration Policy Development
ICRC	International Committee of the Red Cross
IOM	International Organization for Migration
MIFERMA	Société anonyme des mines de fer de Mauritanie
MND	Mouvement national démocratique
MoU	Memorandum of Understanding
OECD	Organization for Economic Cooperation and Development
OMN	Organisation des migrants de Nouadhibou
OMVS	Organisation pour la mise en valeur du fleuve Sénégal
PIRS	Personal Identification and Registration System

PKD	Parti des kahidines de Mauritanie
Polisario	Frente Popular de Liberación de Saguía el Hamra y Río de Oro
PPM	Parti du peuple mauritanien
SEZ	Special Economic Zone
SIJP	Société industrielle de la grande pêche
UGOVF	l'Union des originaires de la vallée du fleuve
UM	Mauritaniain ougiya (currency)
UN	United Nations
UNHCR	United Nations High Commissioner for Refugees
UOMS	l'Union des originaires de la Mauritanie du sud
WAPIS	West Africa Police Information System
WFP	World Food Programme

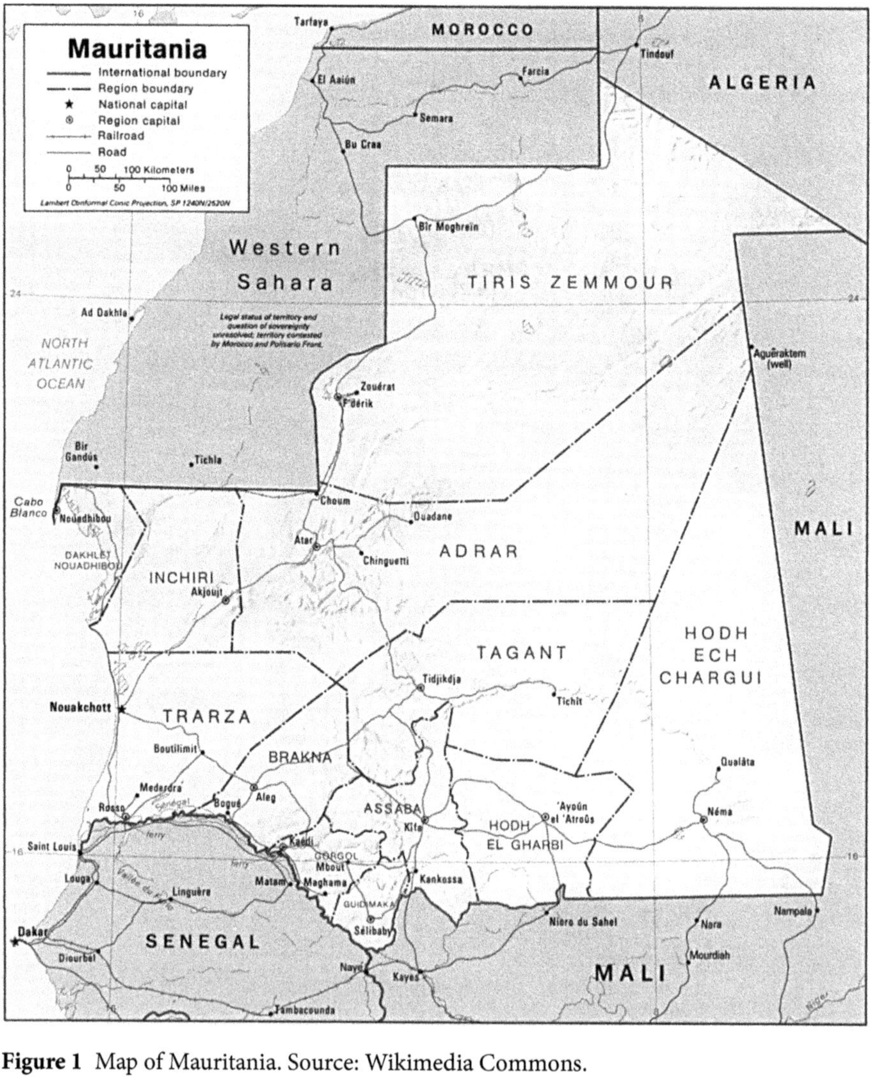

Figure 1 Map of Mauritania. Source: Wikimedia Commons.

Chapter 1

INTRODUCTION

'We could even see the lights of Los Palmos in the distance', Pierre said to me.[1] He was standing by a roundabout next to an informal car wash as he spoke, brandishing a piece of fabric to passing cars in the hope of enticing a customer. Like many others in this site of Nouakchott's urban informal economy, Pierre spends hours each day either sitting in the shade of some trees, waiting for employers looking to hire domestic workers to drop by, or standing at this roundabout, attempting to flag cars whose owners may wish to have them cleaned. The only capital city in the Sahara Desert, Nouakchott's dusty streets and open, glaring skies create a steady demand for car-washing and window-tinting services, which informal migrant workers and others on the urban margins accommodate adeptly. But having only arrived here two months ago from the northern Mauritanian port city of Nouadhibou, where he found himself after being deported from Morocco, Pierre was still learning the tricks of this informal car-washing trade. A fellow migrant worker from Mali who had been on the scene longer, named Khalifa, was showing him the ropes.

Born in Côte d'Ivoire, Pierre travelled to Morocco soon after completing his secondary education in the hope of earning some money for his family. But demoralizing experiences of racism in Morocco soon directed his sights to Europe, which he became set on reaching 'at all costs', he explained, since his family was depending on him. Pierre made two attempts to get to Spain from Morocco. The first was in a boat that set off for the Canary Islands from Dakhla, Western Sahara, only to be intercepted by the Moroccan Royal Marine, Los Palmos glistening on the horizon. The second attempt was at the Spanish-Moroccan land border at Ceuta, where he participated in one of the infamous mass crossings that periodically undermine the European Union's (EU) 'postnational and transborder imaginary' (Stierl, 2020b: 265). On this occasion, Pierre was part of a group of about one hundred that made it over the fences. But in the rush to evade the police waiting for them on the other side, he and a few others became separated from the main group. They hid for a time under a truck, until one member of the group who was parched with thirst called out to a passer-by, who responded by inviting them to shelter in his house with promises of food and drink. But they had been

1. Unless otherwise indicated, all names in this book – be they of migrants, officials, or NGO workers – are pseudonyms.

deceived: 'He said he was just going out to the supermarket, but he came back with the cops!' They were handcuffed and taken at gunpoint from the house.

This was the start of Pierre's long deportation journey to Mauritania. He was unceremoniously deposited by Moroccan security forces in the no-man's land separating Mauritania's northern border from that of the Western Sahara, without so much as an exit stamp on his passport. He then spent ten months in Nouadhibou, before making his way to Nouakchott, where he had been for two months when we met at the car-washing site. 'One year I've found myself here,' he said to me, 'but I never wanted to come.'

Pierre's situation is far from unique amongst those struggling to make ends meet in the informal economy of Mauritania. Peppered throughout this landscape are experiences of border violence and abandonment spanning the length and breadth of the EU's extended Saharan frontier zone (cf. Brachet, 2016: 276). Pierre's Malian car-washing companion, Khalifa, for instance, told me of two unsuccessful attempts to get to Italy from Libya. He eventually travelled to Algeria, from where he was deported to the deserts of northern Niger, making him but one of the 13,000 people the Algerian state abandoned in its southern borderlands over 2017 and 2018, forcing them to cross on foot into Niger, following an increase in EU pressure on North African states to stem migration from their territories (Hinnant, 2018). Tunisian authorities have adopted a similar strategy vis-à-vis Sub-Saharan migrants, hundreds of whom have been abandoned in the Tunisia-Libya borderland in the summer of 2023, around the time of a €100 million deal with the EU dedicated to stemming Europe-bound migration (InfoMigrants, 2023).

The shadow of the EU thus looms large in the stories of Pierre, Khalifa, and so many other migrant workers in Mauritania. Collectively, they are at the receiving end of a calculated, if ad hoc, policy response to the perennial 'problem' of unauthorized migration into European territory.[2] While many solutions have been floated, that around which an unparalleled EU consensus has emerged consists in offshoring and outsourcing migration and border control responsibilities to countries beyond the EU external borderline (Bialasiewicz, 2012). This strategy of 'externalization' has long been a growing element of EU migration and border policy harmonization,[3] but it acquired an unprecedented intensity following

2. The policy and media reaction to sea arrivals and unauthorized border crossings has generally been wholly disproportionate to their statistical significance (Carrera, 2007; De Haas, 2008 (Mainwaring, 2019)), and overlooks the fact that the majority of people residing without documents in Europe overstayed a legal visa rather than crossed a border without one (El Qadim, 2014). Crucially, though, this policy reaction typically creates more of what it claims to want to prevent (Andersson, 2014b; Czaika and Haas, 2014). Hence, the scare quotes.

3. Within the EU context, aspects of the phenomenon go at least as far back as the Barcelona Process of 1995. It was also significantly bolstered off the coasts of West Africa in the mid-2000s, as shall be discussed in more detail in the next section.

the mass arrivals to Europe during the 2015 'summer of migration' (Hess and Kasparek, 2017). This shift was heralded in the 2016 EU-Turkey deal, which effectively traded refugees for a €6 billion cash injection. A similar logic of externalization is evidenced in the deepening of cooperation between the EU and the Libyan Coast Guard from 2017, at a very moment in which evidence of the latter's implication in migrant slave markets and concentration camps in Libya had become incontrovertible.[4] In Niger, a law criminalizing people-smuggling was adopted with EU backing in May 2015, resulting in a decimation of the local transport industry in Agadez and a dispersal of migratory routes into the far more lethal terrain of the Sahara desert (Border Forensics, 2023). Spain and the EU have also equipped and supported the Moroccan security forces that suppressed Pierre's efforts to reach Europe on two separate occasions (Gazzotti, 2021; Vives, 2021). And while formally distinct from the diktats of EU external migration policy, the mass forced migration of Malians from Algeria to which Khalifa was subject nonetheless benefits the EU's efforts to curb unauthorized entries into its territory (cf. Sylla and Cold-Ravnkilde, 2022).

These are but regional examples of what has become a global phenomenon. Across the world, advanced capitalist states have erected increasingly elaborate barriers around their edges and externalized responsibility for the repression of those who dare subvert them (Brown, 2010; Jones, 2011; Walia, 2021). Australia's so-called 'Pacific Solution' involves offshore island detention camps and attempts to broker refugee swap deals with neighbouring states (New Keywords Collective, 2015: 76). Meanwhile, in the borderlands of the United States, migrant deterrence measures can take proactive, violent, and even spectacular form (De Genova, 2013a; Tazzioli and De Genova, 2020), in addition to more passive strategies of abandonment to the biophysical forces of the sea and the desert (Squire, 2017; Stierl and Tazzioli, 2021). In this context too, efforts have been made to strike migration deals with Latin American states and to train and equip their security forces, equally with a view to preventing migration 'at source' (Hiemstra, 2021).

The growing presence of such migration agreements and the proliferation of 'migratory conditionalities' within broader partnership agreements (Gabrielli, 2016) speak to a more holistic set of externalization technologies, of which the EU has also been a pioneer. While entailing a less openly violent approach to the art of migration management at a distance (Cobarrubias, 2019), the founding principle of these 'softer' externalization strategies remains a desire to prevent Europe-bound movement at its perceived point of origin. The Valetta Summit

4. The EU in 2017 increased its operational support to the Libyan Coast Guard and Navy and to the General Administration for Coastal Security (Directorate-General for Neighbourhood and Enlargement Negotiations, 2020). Earlier that year a report by German diplomats described migrant detention centres in the country as having 'concentration camp-like conditions' (The New Arab, 2017). In November 2017, footage of migrants being auctioned in slave markets sparked global outrage, though the practice had been well embedded by that point.

in November of 2015, for instance, gave birth to the EU Emergency Trust Fund for Africa (EUTF),[5] which put forth the slippery concept of the 'root causes of irregular migration'. Skirting the difficulties that inhere in crafting policy around the mist of subjective future intentions, the EUTF established a discursive and financial link between the prevention of 'irregular migration', on the one hand, and a host of broader security concerns and development initiatives on the other. This policy conflation was subsequently dispersed across a range of participant countries, regionally clustered according to a geographic parcelling of the African continent established by the Trust Fund. Within this geographic parcelling, Mauritania is placed in the Lake Chad and Sahel Region, and by virtue of this position availed of €84.4 million worth of Trust Fund project financing in 2019 for the purpose of addressing the 'root causes of irregular migration' (Coopération: Union européenne – Mauritanie, 2020: 20).

From the point of view of migrant workers in Mauritania like Pierre and Khalifa, however, it matters little whether the framework to ensure that they never set foot on European soil is dressed in the language of development or border security. In any case, the infusion of EU migration management prerogatives into a landscape of racialized hierarchies and class polarization has had decidedly negative effects for those on the urban margins. Many at this informal car-washing site recounted experiences of being detained and deported on grounds of lacking a residence permit, and of thus being 'illegal' in the eyes of Mauritanian security forces.[6] This condition of what Nicholas De Genova (2002) memorably termed 'deportability' is a novel one in Mauritania, but it finds ways of reinforcing the existing precarities of the informal economy and exclusions of the prevailing racial order. While the technocratic imperative of preventing 'irregular migration' at source assumes it is possible to identify and capture Europe-bound migrants in isolation, it in practice becomes enmeshed within the global dynamics of capitalist development which are responsible for a situation in which vast swathes of people are left with no other survival option than to offer informal car-washing services on the streets of cities like Nouakchott. And while being deported on the grounds of being 'illegal' is a distinctly new phenomenon in Mauritania, it resonates with older historic episodes in which large numbers of Senegalese nationals and Afro-Mauritanian citizens were deported from state territory as part of a campaign of racial purification, orchestrated by ethno-nationalist elements within the Mauritanian armed forces.

This book is concerned with these areas of intersection between EU external border policy and the social relations and structures that it encounters as it unfolds. Based upon eleven months of fieldwork in Mauritania, it details not only the institutional specificities of the externalization process as it has manifested in this national context (Chapter 4), but also how this manifestation sits in relation to

5. Full name: EU Trust Fund for Stability and Preventing the Root Causes of Displacement and Irregular Migration in Africa.

6. In most such instances, arrests, and detention occur before any formal verification of documents has taken place.

extant racialized social structures and capitalist development processes. To do so, it offers a detailed historical overview of these structures and processes, outlining continuities as well as divergences between the colonial encounter and the era of externalization (Chapters 3 and 4). From this vantage point, it is shown that the racialized basis of belonging within the nation-state today is rooted in governance and territorial shifts enacted during the colonial era. Meanwhile, the book's ethnographic chapters (Chapters 5, 6, and 7) progressively detail how capitalism's tendency to cast people out of the process of its development provides further human material for the EU border regime to illegalize. It is within this coalescence between the wasteful dynamics of contemporary capitalism and a colonially inherited framework of racialized territorial belonging that the EU border regime's illegalization of migrants in Mauritania intervenes.

Despite its activation of these colonial residues, externalization also appears an effort to uphold a bygone colonial dominance in the face of its slow dissolution. To take but one prominent example, the Nigerien anti-smuggling law mentioned above was abrogated in November 2023 by the military junta that took power in Niger earlier that year, part of a regional mobilization against Western interference that has swept the Sahel since 2020 (Niang, 2022). More generally, in an era of what Achelle Mbembé (2021) calls late Eurocentrism, externalization appears far from the omnipotent assertion of power and strength by which the colonial imaginary has long assumed European civilization to be characterized. Within this juncture of waning European dominance there lie epistemic opportunities, ones this book finds in the perspective of Pierre, Khalifa, and others on the urban margins.

*The Mauritanian bridge/*barzakh

The vantage point of Mauritania's social structure and relations holds rich potential for a critical perspective on externalization, and on migration and border regimes more broadly. Since the inception of the Mauritanian nation-state, there has a been a concerted effort to depict this territorial formation as a 'bridge' or 'hyphen'[7] between 'Arab' and 'Black' Africa (cf. Ould Moctar, 2022b). In other words, as an identity construct, the Mauritanian nation is intimately tied to notions commonly associated with borders and boundaries, and in particular the dual quality of separation and connection (Walker, 1999: 447; Mezzadra and Neilson, 2013: 35). This trope has the merit of capturing Mauritania's geographic location as well as its ethno-linguistic composition. Comprising the *ḥassânîya* dialect of Arabic and the African languages of Fula, Soninké, and Wolof, Mauritania's ethno-linguistic profile reaches well beyond the country's territorial borders, encompassing peoples north of the Sahara and south of the Senegal River. Within its borders, these ethno-linguistic categories are represented, on the one hand, by a hegemonic group of *ḥassânîya* speakers, who collectively identify as 'Moors', and on the other by Afro-Mauritanian speakers of Fula, Wolof, and Soninké. The former group is

7. *Trait d'union* in French.

further divided between 'white Moors', or *biḍân*, and formerly enslaved 'black Moors' or *harâtîn*.[8]

It is for these reasons of social composition that Mauritania is often conceived of, in geographic and ethno-linguistic terms, as a point of simultaneous connection and separation (Ahmed Salem, 2004; Pettigrew, 2023). For Zekeria Ahmed Salem (2004), thinking of Mauritania as a 'frontier state' offers insight into both the internal social dynamics between ethno-linguistic groups more commonly thought of in geographic isolation from one another, as well as its corollary external geopolitical position of being simultaneously within and between West Africa and North Africa. Erin Pettigrew (2023), meanwhile, sees in the space of Mauritania a *barzakh*. Originating in the Qu'ran as a concept denoting a bridge between the seen and unseen worlds, the *barzakh* has since been mobilized by Sufi scholars and social scientists alike to theorize an array of spaces and phenomena defined by liminality, dynamic tension, and the dual qualities of separation and connection (Lawrence, 2020; Brower, 2022). For Pettigrew, the *barzakh* serves as both conceptualization of Mauritania's physical location and analytical tool for unveiling the socially constitutive role of the invisible and the divine within the history of its social relations and structures.

Within the more specific domain of migration, Mauritania appears less a bridge between North and West Africa, and more an integrated component of the latter. As we will shortly see, an increase in illegalized migration to the Canary Islands in 2006 recast Mauritania as 'a transit country' from the Eurocentric perspective of the border regime, but this regional migratory embeddedness in West Africa remains the country's primary migratory feature. Unlike its northern neighbours of Algeria, Morocco, and Tunisia, Mauritania's postcolonial relationship to Europe has not taken the form of a substantial outflow of migrant workers to the former colonial metropole. Hein De Haas (2011) estimates that just 20.6 per cent of Mauritanian emigrants reside in North and West Europe compared with 65.6 per cent in the rest of West Africa.[9] Inward migration has been equally regionally oriented, with the majority of the 138,162 foreign nationals estimated to be residing on Mauritanian state territory in 2015 being of Senegalese and Malian origin (UNU-Merit, 2017: 4). It is for this reason that De Haas (2011: 27) observes Mauritania to be 'commonly but misleadingly referred to as a transit country'.[10]

8. *Biḍân* is the word used in the *ḥassânîya* dialect of Arabic to describe the dominant group within the Arabo-Berber populations of Mauritania. It is often translated in French to 'Maure' (Moor). At origin a colonial term, 'Moor' has been thoroughly appropriated colloquially in Mauritania. Nonetheless, given that 'the Moors' are historically quite an ill-defined group, I use the *ḥassânîya* term – *biḍân* – throughout this book for reasons of analytical clarity.

9. This figure in 2015 was recorded to be 23 per cent (UNU-Merit, 2017: 4).

10. And it is for similar but more general reasons that Franck Düvell (2012) argues 'transit migration' to be a 'blurred and politicised concept'.

Of course, these features of Mauritanian society are not static. The history of the country's social formations has involved a passage from precolonial Emirates and kingdoms of porous and ill-defined boundaries to a colonial fixation of territorial and ethno-racial lines, and an ongoing process of postcolonial nation-state consolidation in which these schisms repeatedly manifest. This is, of course, but one manner of periodizing Mauritanian history. Were we to chart the trajectory of capitalist development in the country, we might draw attention to the integration of domestic colonial populations into the international market over the course of the 1920s and 1930s; the subordination of an incipient Mauritanian state within the global capitalist system following the establishment of an international mining operation in the north of the country in 1952; the consequent emergence of an urban industrial working class over the course of the 1960s; and the structural adjustment era's decimation of the formal sector and an associated withering of this class consciousness (Amin, 1973: 79; Bennoune, 1978: 37; Seddon, 1996: 200).

These two processes – the colonial shaping of territorial and social boundaries and the trajectory of capitalist development – will be shown to be of equal significance to the analytical concerns of this book. But as far as periodization is concerned, the emphasis will be on the former. The reasons for this will become clear in Chapter 3, where it will be demonstrated that the original extraterritorial deployment of the border form of spatial organization to Mauritania took place in the colonial era. Moreover, it immediately gave rise to a new bureaucratic approach to managing mobility and an equally unprecedented racialization of territorial belonging, each of which was reinforced following an EU-impelled transformation of the migratory landscape in the mid-2000s.

This arrival of EU bordering prerogatives emerged off the coasts of West Africa in 2006, when the Canary Islands saw upwards of 32,000 unauthorized sea arrivals (Choplin, 2010). As one of the source-points of these unwanted migration flows, the Mauritanian port city of Nouadhibou became a strategic node within an elaborate regional migration policing operation (Carrera, 2007; Casas-Cortes, Cobarrubias and Pickles, 2014). Spearheaded by Spain and funded by the EU, this operation had a logic of pre-emption; by deploying planes, boats, and boots, it was hoped that 'irregular' Europe-bound migration could be prevented at source. In this respect, it set important operational precedents for the outsourcing strategies adopted by the EU post-2015. This was achieved with the aid of Frontex, the European border management agency.[11] Created in 2004 by European Council regulation 2004/7, it would first test its mettle off the Atlantic coasts in 2006, providing expert assistance to authorities in the Canary Islands and logistical support for interceptions at sea.[12] In order to guarantee the offshore implementation of detentions and deportations,

11. Named 'the European Border and Coast Guard Agency' as of 2015.

12. These interceptions would ideally take place within 24 miles of the West African coastline, so that international protection norms could be evaded in Spain and responsibility for detention and deportation would fall exclusively on states such as Mauritania (Carrera, 2007: 21).

Mauritania's first migrant detention centre was quietly constructed in Nouadhibou in April 2006 (Amnesty International, 2008: 22). This soon ushered in a novel form of social despondency in the city, alongside a flurry of intergovernmental and non-governmental activity tasked with managing and mitigating it (Streiff-Fénart, 2011). Thus began Mauritania's integration into the landscape of externalized EU migration and border governance, a role that successive governments in the country have by and large enthusiastically sustained.

In detailing this encounter between EU border externalization and the socio-historic landscape in which it unfolds in Mauritania, this book necessarily confronts the issue of Eurocentrism. Indeed, insofar as it consists in marrying the highly mediatized issue of EU external border policy to the rather more obscure topic of social relations and structures in Mauritania, it directly addresses the 'asymmetric ignorance' by which Dipesh Chakrabarty (2000: 28) once characterized Eurocentric thought. For this reason, we ought to momentarily dwell on what is at stake in overcoming the phenomenon of Eurocentrism, both in the study of the EU border regime and in more general terms.

After Eurocentrism

It should come as no surprise that the EU border regime operates on assumptions that are necessarily Eurocentric (cf. Casas-Cortes and Cobarrubias, 2019). In seeking to prevent Europe-bound migration at source, it reduces vast proportions of humanity to nothing more than 'potential candidates for irregular migration', projecting on them aspirations that often have little to do with people's subjective desires, while also flattening out rich and complex social landscapes, rendering them mere source-points of unwanted Europe-bound movement. This book overcomes this reductive, two-dimensional perspective on the historical and social landscapes in which EU external border policies operate. At an empirical level, it does so by expanding the analytical lens beyond those who wish to reach Europe (cf. Newhouse, 2018), instead examining how EU bordering practices and logics find them where they are. Even with the empirical lens so expanded, however, a theoretical challenge remains; how to afford the externalization process the detailed analytical attention it requires, without absorbing its own Eurocentric conceptual assumptions?

In a sense, this challenge is simply that of overcoming the old spectre of methodological nationalism (Wimmer and Glick Schiller, 2002), but it takes unique form and urgency for critical analyses of contemporary migration and border regimes (De Genova, 2013b). Indeed, within the analysis of the global politics of migration, the normative framework of 'Western' liberal capitalist democracy carries disproportionate analytical weight (Walters, 2015: 11; Korvensyrjä, 2017), resulting in parochial and contingent assumptions masquerading as global ones (Berriane, de Haas and Natter, 2015; Vigneswaran and Quirk, 2015). These normative assumptions are, moreover, heavily inflected with the legacy of the colonial era, which has left an indelible mark on the discipline of migration studies (Mayblin and Turner, 2021). To offset these blind spots, and to avoid 'the superimposition of Euro-centred categories and narratives onto any landscape of mobility' (Garelli and Tazzioli, 2013: 247), calls have recently been made to analyse externalization from the Global South (Stock, Üstübici and Schultz, 2019), to consider the agency and room for manoeuvre of non-EU state actors in the negotiation of externalization (El Qadim, 2014; Cassarino, 2018), and to reflect on how border regimes are conditioned by colonial power dynamics and legacies (Brambilla, 2014; Lemberg-Pedersen, 2019; Stierl, 2019; Lemberg-Pedersen et al., 2022).

By foregrounding the socio-historic landscape in which the EU border regime intervenes in the Global South context of Mauritania, this book draws from and contributes to these perspectives. As the third and fourth chapters illustrate, the form that externalization takes is indeed conditioned by colonial-era transformations in territorial organization. At the same time, however, the book outlines the nature and scope of the agency of a range of actors in Mauritania. These include state actors, who find ways of exercising leverage over Spain and the EU despite the asymmetric relationship in which migration control negotiations occur, as well as international and national NGOs in Mauritania, who are faced with financial and social opportunities as well as political compromises within the new landscape of externalization in the country (Chapters 5 and 7). Most importantly, it foregrounds the perspective of illegalized migrants themselves, such as that of Pierre, Khalifa, and many others. Collectively, these elements help drive an analytical wedge between the Eurocentric assumptions of the border regime and its critical analysis.

The question of moving beyond Eurocentrism is not, however, simply one of disavowing the conceits of the EU border regime. Rather, it goes to the heart of the ideological underpinnings of the modern world system, and the Gramscian interregnum in which it currently finds itself.[13] For, on the one hand, the fabric of the modern world system was woven by the colonial encounter (Quijano, 2000; Mignolo, 2012), whose asymmetric imprint is epitomized in the relationship between Europe and Africa. This is evidenced in the uneven development between the two continents (Amin and Girvan, 1973; Rodney, 1973), in the stance of international legal institutions vis-à-vis African states (Ba, 2017), and in the colonial structures underpinning local government administration and social welfare policies in Africa (Mamdani, 1996; Mkandawire, 2016). On the other hand, however, evermore glaring cracks are emerging between these dynamics of the world system and an archaic assumption of European civilizational superiority, whose attendant economic and institutional orderings nonetheless continue to shape it. Within these cracks, geopolitical alliances shift, new ones emerge, and the agency of formerly colonized African states more forcefully asserts itself, as recent events in the Sahel suggest (Brown, 2012; Chipaike and Knowledge, 2018).

13. Whereby the old is dying and the new cannot yet be born.

Situated within the paradoxes of this historic interregnum, the externalization process appears to be at once a slow dissolution and a beleaguered continuation of much more deep-seated structures. These structures have their roots in a fifteenth-century systemic shift, which saw the centre of gravity for commerce and exchange slowly move away from the Mediterranean towards the Atlantic (Amin and Girvan, 1973). What followed has been an era of unprecedented European dominance, operationalized in the global spread of capitalism and the linear form of territorial division (Anievas and Nişancıoğlu, 2015). In tandem, a Eurocentric ideological construct emerged to rationalize and mask the mass violence, exploitation, and enslavement by which these material transformations were characterized (Amin, 1988). By the time of the French colonization of West Africa at the turn of the nineteenth century, the asymmetric relations between the European and African continents that emerged from this shift had been well and truly entrenched, as had the teleological conceits of the faux universalism espoused within its Eurocentric imaginary.

Today, this Eurocentric imaginary is embodied in the European Union, whose origins are inextricably bound up in the colonial heritage of its founding member states (Hansen and Jonsson, 2014).[14] As a result, the EU's very sense of ontological selfhood and security today depends upon its asymmetric relationship to Africa (Haastrup et al., 2021). But at the same time, this asymmetry is increasingly faltering in the face of a protracted and multifaceted crisis of both European and global dimensions.[15] Failing any internal resolution to this crisis, and in the absence of any interrogation of the colonial origins of the EU's contemporary sense of ontological security, externalization emerges as a last-ditch, drawn-out effort to keep colonial asymmetries between Europe and Africa and alive, in the face of their slow dissolution.[16]

14. At the time of the first steps towards European integration, in the form of the Schuman Declaration and the 1957 Treaty of Rome, it was taken for granted that integration would entail shared management of British and French colonial territories (Hansen and Jonsson, 2014: 124). Moreover, with both colonial powers facing losses in Asia and the Middle East at this time, Africa was viewed as the key to a rejuvenated and 'communitarized' European colonial project.

15. A non-exhaustive overview might include a sovereign debt crisis, the Brexit referendum, a resurgent far-right, a global pandemic, the Russian invasion of Ukraine, and an inflationary spiral sparked by the latter two crises.

16. Interestingly, nowhere is the colonial asymmetry of the Eurafrica project detailed by Hansen and Jonsson more stark than in the domain of migration. At the time of its intellectual birth in the interwar period, it was assumed as natural that Africans should not be permitted in Europe, as this would result in racial impurification. At the same time, in addition to being a well of untapped resources, Africa was viewed as an emigration safety valve for European populations, who would, naturally, bring morality, culture, and civilization to the continent (2014: 38). While it would be unthinkable for an EU politician today to speak in such openly racist terms, the unprecedented restrictions on safe and legal avenues to migrate from Africa to Europe indicate that the logic has been thoroughly structurally absorbed.

The morbid symptoms of this interregnum are manifest, particularly for Pierre, Khalifa, and others at the receiving end of Eurocentrism's death throes. But light can also be glimpsed through the cracks. Indeed, at the twilight of Eurocentrism, there is greater room to explore what Robbie Shilliam (2011) once described as the 'perilous but unavoidable terrain of the non-West'. As this book demonstrates, within the collective subjectivity of those facing the violent triad of economic abandonment, racialization, and illegalization in West Africa, there lie intuitive glimmers of a more encompassing universalism than that which has historically served to rationalize and cloak the violence and exploitation of the modern world system. These are brought to light in this book through a concluding ethnographic exploration of the meaning-making strategies of those on the urban margins in Mauritania. As we will see, theirs is a perspective that foreshadows a universalism unshackled from the Eurocentrism that undergirds the externalization process.

To speak of an 'after border externalization' thus carries three levels of meaning. At the most empirical, it refers simply to what happens after the externalization process has been introduced in a given context. How does the policy prerogative of policing 'irregular migration', for instance, interact with existing racial hierarchies and economic relations? As we will see, to answer such questions it is necessary to also look to a time before externalization, to chart the genesis of the relations and structures in which it intervenes. At a more general level, 'after border externalization' is also an acknowledgement of scholarly reflections on the need to move beyond the term 'border externalization' (Gross-Wyrtzen and Gazzotti, 2021; Frowd, 2022), the logic being that it is Eurocentric and captures relatively little of how transnational security interventions concretely unfold (cf. Raineri, 2021). This book is sympathetic to these claims, and for this reason at no point considers externalization in isolation from broader dynamics and processes, and indeed spends quite a bit of time discussing things that have little on the surface to do with externalization. At the same time, however, it holds that there is something of central significance in the relentless and seemingly irrational drive to outsource migration controls. This speaks to the third level of meaning at play in this book. Externalization is indeed a particularly visible material expression of Eurocentrism. But more precisely, it is an expression of the decline of Eurocentrism, of Europe's inability to live up to its proclaimed self-image (Stierl, 2020b). To speak of an 'after border externalization', then, is to speak of a time after Eurocentrism. While detailing the material dynamics of this juncture, the book also examines its epistemic openings.

On method and fieldwork

The research upon which this book is based was carried out over the course of eleven months spent in Mauritania in 2017 and 2018. In order to methodologically foreground the socio-historic landscape in which the border regime intervenes, it adopted a multi-sited ethnographic approach (Marcus, 1995), which is concerned with integrating situated, place-specific analysis and world system dynamics. To this end, three urban locales in Mauritania served as sites of investigation, namely the northern port city of Nouadhibou, the southern border town of Rosso, and the capital city of Nouakchott. A chapter is devoted to analysing the externalization process in each of these locales. As shall be illustrated, each urban centre researched represents a slightly different expression of how the externalization process relates to the Mauritanian border and its social relations and structures. As a consequence, the border regime's illegalization of migrant workers takes distinct form in each locale. For this reason, the multi-sited ethnographic approach proved fruitful to investigating this intersection between these place-specific social relations and the larger systemic forces of the border regime.

Research across these spatial sites involved interviews and ethnographic investigations within three overlapping, but distinct, social worlds in Mauritania.[17] The first was the level of officialdom, which encompasses the diverse and contradictory range of actors who have a common stake in the newly constituted domain of migration and border management in the country. The most prominent of these are the European External Action Service in Mauritania (EEAS) and the International Organization for Migration (IOM). Their closest points of national contact are based in the Ministry of Interior which, alongside the Ministry of Defence, carries the greatest degree of national institutional responsibility in the domain of migration and border management (Mohamedou, 2016). Semi-structured interviews with officials within each of these bodies offered insights into the concrete workings and malfunctions of EU external border policy in the country, making them key to fleshing out the observations in the literature on how externalization is appropriated, distorted, and manipulated in countries to which migration management prerogatives are externalized (El Qadim, 2014; Cassarino, 2018; Lemberg-Pedersen, 2019).

Alongside officialdom, I held numerous semi-structured interviews and far more informal conversations with members of international and national civil society organizations in Mauritania. Most of the Mauritanian organizations represented here predate the era of externalization,[18] but they have taken migration up as a sphere of concern since the dawn of this era. Many of the international NGOs whose representatives I interviewed had similarly recalibrated their activities in light of the externalization process.[19] Beyond these broad-based organizations, interviews were also conducted with members of the numerous migrant community associations that are active in Mauritania. These are variously organized along sub-national, national, and regional lines. At the regional scale, I engaged frequently with the *Cadre de Concertation des Migrants en Mauritanie*

17. In all instances, the SOAS Code of Practice on Using Personal Data in Research was adhered to, in ways tailored to the particular social sphere of investigation.

18. These include the AMDH (*Association mauritanienne des droits de l'homme)*, APEAH (*Association pour l'environnement et l'action humanitaire*), AFCF (*Association des femmes chefs de famille*), and the FONADH (*Forum national des ONG de droits de l'homme*),

19. The example of the Red Crescent is most noticeable in this regard due to its role in the management of *Guantanamito*, the Nouadhibou detention centre.

(CCMM), an organization representing numerous West African communities in Nouakchott. In addition to interviewing its convenor and various participants, I was invited to sit in on their monthly meetings. These exchanges offered insights into how migrant community associations themselves shape and are shaped by the externalization process, while also facilitating a bird's-eye perspective into how it has impacted their communities.

At the same time, much is obscured at the exclusive level of the bird's-eye view. Indeed, despite the overwhelming majority of migrants in Mauritania being employed in the informal sector, their subjective experience and input rarely figure within formalized spaces of overlap between civil society and officialdom. For this reason, the informal economy of Mauritania constituted a vast portion of the ethnographic research upon which the book is based. In practice, this entailed 'deep hanging out' (Madison, 2005) at various labour pick-up points in the three cities in which this research took place and getting to know people. For the most part, these were individuals who float in and out of precarious employment as bricklayers, street vendors, electricians, plumbers, domestic workers, cooks, car washers, and other such activities that take place beyond the reach of state regulation. Through this approach, I got to know upwards of eighty individuals at various sites of the informal economy in Nouakchott, Nouadhibou, and Rosso over the course of fieldwork. The duration of these encounters and relationships varies from fifteen or twenty minutes to the eleven months I spent in the field. All, however, entailed at least a mutual exchange of basic information (name, nationality) and a sharing of relevant details about what I and my interlocutor were doing in Mauritania.

Of course, the nature of my interaction with each of these social realms was irreducibly conditioned by own subjective experience and structural position (cf. Altorki and El-Solh, 1988; Mullings, 1999; Ganga and Scott, 2006). Being the European son of a Mauritanian *biḍân* was a determining factor of how I was received in the realm of officialdom, but it was not a sure predictor of how this reception would unfold. In certain instances, interviews with state officials could be prefaced with a discussion of familial background, while in other instances, the screening questions were limited to topics pertaining to my research project. As regards civil society, I had already carried out a research internship with the *Association mauritanienne des droits de l'homme* (AMDH) in 2015, which meant that I could draw on pre-existing links at the time of my commencing this research. It was in my capacity as a former intern and sympathetic researcher willing to lend my skills that I was invited to attend CCM meetings, which were held in the offices of the AMDH.

The process of navigating the informal economy was – unsurprisingly – more informal, but for this reason, I sought to uphold a stringent interpretation of informed consent, respect for boundaries, and the individual's right to privacy. Given the lack of formal education and low literacy level of many in the informal economy,[20] the words 'doctoral research' and 'PhD' did not always register. I

20. Of course, the lack of formal education in no way precludes impressive levels of informal education and intellect. Most people I interacted with spoke anywhere between three and six languages, and appeared to have little difficulty in picking up the language of their environment.

therefore often simply explained that I was writing a book about migration, and was interested in people's experiences. During such encounters, I did my utmost to be attuned to the slightest non-verbal signals that might indicate a person's unwillingness to be part of an exchange. Being so attuned seemed essential given the structural inequities in which the relationship between researchers and research participants takes place. These latent power dynamics could well have meant an individual might feel reluctant to voice their discomfort, even if it were subjectively felt. For this reason, I always tried to sense any such potential discomfort and move on at the slightest intuitive hint. While intuition may have therefore led me away from people unnecessarily on occasion, exercising a precautionary principle in this regard seemed more important than gathering data.

Of course, the partial standpoint upon which this research – like all research – is based means that gaps and oversight inevitably emerge. The nature of such omissions was most clearly articulated by some of my companions in the informal economy, as illustrated in the following excerpt from fieldnotes. After speaking of the 'important' work that I was doing, one Afro-Mauritanian offered the following advice: '"If you don't work with the people, if you don't get inside, you're not going to understand because … because …" He hesitated, searching for the words, despite seemingly having already found them. "Because you're not inside?" I ventured. "Exactly."' The interior referred to here is most obviously delimited by the very different livelihoods by which I and my interlocutors in the informal economy get by. But it is also further interlaced with ethno-linguistic and racial boundaries. Upon initially seeing me, for instance, individuals in the various sites of the informal economy would invariably assume I was a Mauritanian or Arab national looking to hire workers. Undoing this impression and establishing a different dynamic was of utmost ethical and methodological importance.

Even when headway had been made in this regard, however, linguistic barriers could occasionally still manifest. All of my interactions took place either in French or the Mauritanian *ḥassânîya* dialect of Arabic. Given that these are the effective *lingua franca* of the multi-ethnic informal economy of Mauritania, this allowed for a good deal of communication and understanding of what was being said around me. But my lack of any basic understanding of other prominent languages, such as Soninké and Fula, meant that I did not always follow conversations within the smaller social groups that would occasionally form along ethno-linguistic lines throughout a given day.

Another related limitation is the relative lack of women's experience represented in this book. If, as mentioned previously, attempting to redress the latent power disparities of being a 'Western' researcher navigating the informal economy already proved a delicate enough exercise, it became all the more so when gender dynamics were involved. My aforementioned desire to respect personal boundaries and be attuned to non-verbal cues was all the more palpable where women in the informal economy were concerned. The consequent relative absence of women from the narrative is somewhat compensated by interviews with certain women heads of community associations, but these do not attain the same level of experiential representation as that of men within the ethnographic portions of the book. While there is a growing wealth of scholarship on the gendered dimensions of the EU border regime in the region and the particularities of women's experience of externalization (Freedman, 2012; Gonzalez, 2012; Stock, 2012; Tyszler, 2019), I hope that someone more qualified than myself (ideally, a Mauritanian woman) can highlight all that is missing in this regard from this analysis of the Mauritanian context.

The field notes upon which the ethnographic chapters of this book are based therefore attempt to speak across a vast structural chasm forged by class, race, gender, and citizenship. The existence of this structural chasm poses ethical and political questions for social science research, and for migration research in particular (Anderson and Ruhs, 2010; Düvell, Triandafyllidou and Vollmer, 2010). Indeed, given the noxious state of migration politics across the globe, the question of how to prevent research findings being used for the purpose of what are often lethal schemes of governmental control ought to be paramount (Pezzani and Heller, 2013: 293; Stierl, 2020a). As far as this research process was concerned, this question was primarily addressed by ensuring that information gathered during ethnographic investigations was kept separate from the realm of officialdom.

Beyond this, I made regular feeble attempts to mitigate this structural chasm and offset the necessarily extractive nature of fieldwork in Africa (Ba, 2022). In practice, this meant things like bringing gifts to companions in the informal economy – cigarettes, English lessons, packets of tea leaves to make mint tea – or, when necessary, helping hide the materials of street vendors and workers before they could be confiscated by municipality officials. I also offered verbal and written translation services to the AMDH on occasion and alerted them to potential cases of concern encountered during ethnographic investigations. As regards research dissemination, beyond anonymization and data protection, there is further need to consider whether the content might inadvertently mirror the prerogatives of the border regime itself, by shining a spotlight on illegalized migrants (De Genova, 2002; Pezzani and Heller, 2013). In this light, this book's analytical desire to situate this illegalization within pre-existing social relations and structures in Mauritania is an ethical imperative, as well as an analytical one.

Ultimately, however, no mere piece of academic research can undo the violent material inequalities that are examined in this book, nor rearrange the vast structural chasm separating those encountered in its pages from me and many of its readers. While the shared experience of fasting Ramadan meant I and my companions in the informal economy fleetingly occupied a similar visceral and spiritual terrain while I was on fieldwork,[21] the ethnographic narrative should not be read as an attempted 'emic' perspective on the essence of people's lives,

21. Fasting, I would tentatively submit, heightens one's capacity for empathy and understanding. But in any case, it certainly provided plenty of openings for casual exchange about hunger and thirst in the hot and dusty conditions of Nouakchott in May.

and much less an objectivist account of things as they 'really are'. Rather, the focus throughout is on the empirical details and theoretical implications of the encounter between the EU border regime and historically endowed social relations, structures, and subject positions in Mauritania, all seen from my own necessarily partial standpoint.

Chapter outline

This analytical goal is developed in the book over three parts, which do not include this introduction and the conclusion. The first part, consisting of one conceptual chapter, develops a framework for analysing the EU border regime in a non-Eurocentric fashion. It first defines Eurocentrism with reference to the origins of the term, before discussing the potential for a non-Eurocentric universalism in the present juncture compared with the moment of these origins. It then reviews existing literature on the externalization process, drawing from it two analytical strategies for overcoming Eurocentrism in the study of externalization. In brief, these consist of historicizing the border regime's external operations and situating the illegalization of migration involved in these operations within pre-existing social relations and structures. The chapter then offers an overview of what this historicization looks like in general terms, situating the EU external border within a deeper history of European supremacy, one spanning five centuries and two waves of colonialism and which, it is argued, is reaching its endpoint with the externalization process. It then locates the border regime's illegalization of migrant labour within the social relations and structural processes of the Global South contexts in which externalization policies are typically implemented, adopting the standpoint of the illegalized to examine these relations and processes.

The second part of the book (Chapters 3 and 4) deploys the historical dimension of this conceptual framework to detail how border externalization fits within colonial and postcolonial history in Mauritania. Chapter 3 details a qualitative shift in territorial organization that occurred in the colonial era. This involved an externalization of the border form of territorial organization from the French metropole to West Africa, which, in turn, spawned new bureaucratic strategies of managing human mobility and an equally novel racialization of territorial belonging. Mauritania was, moreover, a key axis along which the fifteenth-century world systemic shift from the Mediterranean to the Atlantic that birthed the era of Eurocentrism occurred. Chapter 4 retains a focus on the centre-periphery dynamics that emerged from this shift, albeit in the form of the externalization process in Mauritania. In this guise, however, the European dominance that characterized the colonial and early postcolonial periods appears to be in decline. The colonial shift in territorial organization is shown to condition the form that externalization takes in the country, but there has also been a rearrangement of Mauritania's positioning within the centre-periphery imaginary of the externalization process, one that signals a contraction of this imaginary.

The third part (Chapters 5, 6, and 7) adopts the standpoint of migrants in Mauritania to situate their illegalization within the social relations that have emerged from the history detailed in the preceding two chapters. Chapter 5 brings us to the northern port city of Nouadhibou, whose recent conversion into a flashpoint of the EU border regime has marked a departure from its historic role as pole of regional migration. While this early postcolonial period in the city was characterized by the exclusive dominance of European capital, trade and investment flows today are more diverse in their origin and destination. Nonetheless, these flows are facilitated by the EU border regime's illegalization of a precarious and transient Europe-bound informal migrant workforce in the city. It eventually becomes clear, however, that an exclusive focus on such Europe-bound trajectories obscures deeper elements of Mauritania's postcolonial trajectory.

For this reason, the sixth chapter moves to the southern border town of Rosso. By virtue of straddling the Senegal River – and thus the border between Senegal and Mauritania – Rosso offers a unique insight into the living contradictions associated with the original colonial deployment of the border form of territorial distinction to Mauritania. This comes to light through ethnographic examinations of the Rosso border crossing, and discussion of the experiences of migrant workers in the town who were not on the move to Europe when I met them. From this perspective, the border regime's imbrication within a regional history of racialized displacement and dispossession comes to the surface. So too does a collective subjectivity grounded in this regional history, whose perspective on the Rosso border is presented in the final section of this chapter.

At the same time, however, the constitutive and agential role of illegalized migrants vis-à-vis the border regime is not immediately apparent from the vantage point of Rosso's particular regional history. The final substantive chapter adopts a more global frame of reference by moving inward from the borders to the capital city of Nouakchott. Here, the experiences of illegalization and violence suffered by Pierre, Khalifa, and others in the urban informal economy are located within a more general subject position, one shared by migrant workers and Mauritanian nationals alike. Only by foregrounding the agency of migrants does it become possible to situate the border regime within this broader structure of spatially managing capitalism's racialized outcasts.

That said, the discussion at this register remains somewhat aloof from this agency. Chapter 7 accordingly concludes with an exploration of how members of this social group understand and interpret their own agency. This foray leads us away from the analytical tools that have guided the previous two parts of the book. But it also highlights the presence within this collective subjectivity of an intuitive sense of the non-Eurocentric universalism discussed at the start of the first chapter. The conclusion marries these two aspects – the meta-theoretical and the most empirically derived – to tentatively explore what our intellectual and political landscape could look like in the wake of Eurocentrism.

Chapter 2

THE SUM OF ITS PARTS?

Depth, breadth, and perspective in the analysis of externalization

Over the course of the 1190s, the renowned Sufi scholar Ibn 'Arabi made a number of trips from his hometown of Seville, Southern Spain, to the north of Africa. Moving between Tunis, Fez, and Seville, he forged ties with fellow scholars, poets, and mystics in his travels (Austin, 1980). These exchanges birthed enduring insights into questions of unity and differentiation, the relationship between subject and object, and humanity's link to the cosmos. Through prolific and sophisticated writing, he offered, in other words, a penetrating perspective into what Sandro Mezzadra and Breit Neilson (2013: 95) describe as 'the perennial question of the many and the one'. Their opacity notwithstanding, these insights would far outlive the context of their origins. Indeed, despite being separated by more than 700 years, Ibn 'Arabi's thought figured prominently in the worldview of nineteenth-century anti-colonial leaders fighting French encroachment in Africa, such as the Algerian 'Abd al-Qādir al-Jazā'iri, and the Mauritanian Shayk Mâ' al-'Aynâyn, whom we will encounter in the next chapter (Chodkiewicz, 1991: 6; Blalack, 2020b).

Spanning Southern Spain, Northern Morocco, and Tunisia, the West Mediterranean through which Ibn 'Arabi travelled differed markedly from its present-day political and symbolic configuration. Indeed, the contemporary West Mediterranean stands divided, the site of a lateral split between its European and African portions. Of course, this split is physically reinforced by the militarized EU external border that separates the Spanish enclaves of Ceuta and Melilla from Morocco. But there is also a less tangible symbolic divide at play. With the era of European dominance and conquest that emerged at the turn of the fifteenth century, the Mediterranean came to be seen as an existential boundary separating Europe from its 'outside'. Within what Walter Mignolo (2012) calls the imaginary of the modern world system, it became a threshold beyond which Europe was to be defined against its various Others (Amin, 1988: 15). When Ibn 'Arabi made his voyage from Seville to Fez, on the other hand, there was no such existential barrier separating these two locations. On the contrary, they had been unified from the eleventh century under the Almoravid Empire, whose territorial purview stretched northwards from central Morocco across the Mediterranean into Southern Spain, and southwards into the Sahara, throughout much of Mauritania and eastern Mali (Pettigrew, 2019).[1]

What do border externalization policies have to do with this shifting geography of the Mediterranean? How does the outsourcing of migration control policies to African states relate to the history of European dominance that transformed the Mediterranean into an imaginary threshold of civilization? And what can the vantage point of Mauritania, situated at the margins of this geographic imaginary, tell us about all of this? This chapter addresses these question by conceptualizing the relationship between the externalization process, the social relations and structures of the contexts of its implementation, and the Eurocentric imaginary of the Mediterranean. As EU migration and border policies are relocated to African contexts, the illegalization of migration that ensues invariably intersects with pre-existing social relations and structures. Analysing this encounter requires a distance between critical scholarship and the assumptions of the EU border regime, which are, naturally, Eurocentric in nature (cf. Casas-Cortes and Cobarrubias, 2019). Forging this critical distance, in turn, shows how this encounter sits within a broader juncture marked by a relative waning in exclusive European dominance (cf. Mbembé, 2021), a decline reflected in the contemporary standing of the Eurocentric geographic imaginary of the Mediterranean.

To develop these points, this chapter first stakes out the term 'Eurocentrism' with more precision, exploring what it meant in its inception and detailing how this interpretation relates to more recent scholarship, as well as to the contemporary juncture of European decline. We then turn to existing literature on border externalization, drawing from it certain analytical strategies for driving a wedge between the Eurocentrism of this process and its analysis. In a word, these consist of historicizing externalization and situating the illegalized subjectivities it creates within broader sets of relations and structures. The rest of the chapter addresses each of these objectives in turn. Turning first to history, the second section examines the material and symbolic role of the Mediterranean within a broad arc of European supremacy and dominance. The more recent history of colonial rule in Africa is also examined, as it has a direct bearing upon the material form externalization takes in its implementation. In each respect, externalization appears a last-ditch effort to uphold this colonial dominance in the face of its slow decline. The final section moves on to the condition of migrant illegality that externalization creates in its unfolding, situating it within a more general subject position of living labour. From this perspective, the social relations and structures that facilitate the border regime's illegalization of migrants come to the fore.

Together, these strategies of analysing externalization with greater temporal depth and social breadth overcome the Eurocentrism of the process itself. In privileging the perspective of those at the receiving end of externalization, however, they also offer empirical means of grasping the epistemic opportunities that lie in

1. Almoravid dominance gave way to that of the Almohads toward the end of twelfth century, but a similar territorial contiguity remained.

the decline of Eurocentrism. To acquire the conceptual grounds of this endeavour, let's first consider how externalization relates to the broader phenomenon of Eurocentrism, of which it is a beleaguered expression.

Drift, decline, renewal

As literature on EU border externalization has developed in tandem with the phenomenon itself, critical scholarship has become preoccupied with avoiding the Eurocentric assumptions of the border regime.[2] For Aino Korvensyrjä (2017), this means addressing the broader 'Westphalian imaginary' of critical migration and border studies, which often overlooks the colonial histories of violence prolonged by contemporary migration regimes. In a similar vein, Lucy Mayblin and Joe Turner (2021) highlight an affinity between the discipline of migration studies and colonialism (cf. Lemberg-Pedersen et al., 2022). As regards analysis of externalization, this affinity can take the form of an ethnocentrism associated with nineteenth-century European thought, which depicts the behaviours of non-European actors as inevitably wayward and irrational (El Qadim and İşleyen in Cobarrubias et al., 2023). Within both the study of externalization and the broader discipline of critical migration studies, then, there has been a tendency to unthinkingly reproduce epistemic frames and assumptions of European thought, ones often wedded to its colonial past (New Keywords Collective, 2015; Gross-Wyrtzen and Gazzotti, 2021). Such considerations require stepping outside the discipline of migration and border studies strictly defined, and interrogating what is more broadly at stake in this term. In what follows, we first explore what Eurocentrism meant at its conception, and how this understanding fares in the contemporary juncture. We then turn to the literature on EU border externalization, drawing from it two analytical strategies for overcoming Eurocentrism in the analysis of externalization, which inform the rest of the book.

Eurocentrism, universalism, and the Oneness of Being

The question of Eurocentrism – the form that it takes, the strategies for overcoming it, and indeed whether it should be overcome – has long stimulated, and at times hindered, debate within sociology, international relations, and development studies (cf. Gülalp, 1998; Bhambra, 2007; Shilliam, 2011; Anievas and Nişancıoğlu, 2015; Wai, 2018). As an intellectual practice, Eurocentrism is premised upon the assumption of an ontological distinction between European history, society, culture, and experience, on the one hand, and those of the rest of the world on

2. Terms such as 'methodological Eurocentrism' and 'methodological Europeanism' each give voice to this concern (Garelli and Tazzioli, 2013; Van Baar, 2016), as do calls to provincialize Europe in the study of international migration (Walters, 2015; cf. Berriane, de Haas and Natter, 2015: 505).

the other (Sabaratnam, 2017: 20). This assumption is just that, however, and as such, like most ontological assumptions, is rarely subject to critical scrutiny by its holders. Instead, there appears what Walter Mignolo (2002: 65) calls 'the absent location of thinking' – the tendency to think within the particular spatial and temporal confines of modern Europe, and from here present the conclusions that emerge as universal, generalizable ones. In this manner, Europe becomes 'the sovereign theoretical subject of all histories' (Chakrabarty, 2001: 27), a permanent yardstick against which the condition and progress of 'the Rest' are to be judged.

In making such observations, however, we would do well to avoid endowing Eurocentrism with the same mystifying, metaphysical quality with which it itself shrouds the history of European dominance and conquest. When he coined the term 'Eurocentrism', Samir Amin (1988, cf. Amin, 2009) sought to highlight an artificial ideological construct that masks the violent origins of the world capitalist system as well as its exploitative dynamics of uneven development in the present. In their place, Eurocentrism fabricates a mythic continuity between the society and culture of Greek and Roman Antiquity and those of the Renaissance. This serves to naturalize the enslavement, exploitation, and inequality of capitalism's global expansion, presenting them as inevitable correlates of a European civilizational superiority that has existed since time immemorial. Given the fifteen centuries of human experience separating Classical Greece from the Renaissance, the construction of this continuity requires momentous erasures, ones Amin meticulously unearths in the first half of the book.[3] In doing so, he unveils the historically specific circumstances in which Eurocentric ideology emerged, namely 'the crystallization of capitalist society in Europe and the European conquest of the world' (Amin, 1988: 71). Far from being an enduring metaphysical fact, then, Eurocentric dogma emerged from a particular set of historical circumstances, which makes it liable to dissipate or collapse entirely as circumstances evolve, as we will later see.

That said, Eurocentrism has acquired a force of authority that makes it particularly resilient in the face of evolving political trends and intellectual currents. For Amin, much of this ideological resilience lies in a false, provincialist universalism espoused by the Eurocentric intellectual project, which serves as a fig leaf masking the violence and inequality of the world capitalist system. Key to this Eurocentric universalism is an axiom that holds that 'the only possible future for the world is its progressive Europeanization' (Amin, 1988: 107), an assumption that has long been embraced within the domain of EU external relations, as shall soon be discussed. While Amin envisioned a genuine universalist horizon beyond this Eurocentric variant, he did not himself elaborate it: 'I am not attempting to propose a complete formulation of a system to replace the current one. I simply hope that the reflections proposed here will constitute a useful contribution to the elaboration of a universalism liberated from the limits of Eurocentrism' (Amin, 1988: 116).

3. These erasures also make racism a crucial element of this enterprise, which comes to the surface whenever the limits of Enlightenment universalism are reached (Amin, 1988: 94).

Amin's reservations in this regard are not surprising, as 'a universalism liberated from the limits of Eurocentrism' necessitates a foray into what Robbie Shilliam (2011) has called the 'perilous but unavoidable terrain of the non-West'. For Shilliam, this terrain is a perilous one because the very binary of 'West' and 'non-West' is itself an artificial colonial construction. To draw upon 'non-Western' thought as if it was an *a priori* category is thus to reproduce an 'inverted Eurocentrism' against which Amin himself warned (Amin, 1988: 78; cf. Lazarus, 2002: 51). But at the same time, this terrain is unavoidable because otherwise, one risks falling for a 'fallacy of composition whereby an idealized Western modernity is mistaken for global modernity' (Shilliam, 2011: 15). In other words, one risks reproducing the same Eurocentric universalism that Amin derided as a provincialist cover for the inequalities produced through global capitalism.

In the thought of Ibn 'Arabi, there is an opportunity to avoid these perils of 'non-Western' thought while embracing its inevitability. Indeed, his is a universalist ethos that sits outside the false universalism of the Eurocentric project. This universalism is rooted in his concept of the Oneness of Being. For Ibn 'Arabi, the fabric of reality is defined by an indissoluble unity: 'in existence there is no similarity or dissimilarity, for there is but One Reality, and a thing is not the opposite of itself' (Ibn 'Arabi, 1980: 108). Within this singular reality, any seemingly individual entity is but a refraction of a larger whole – the Oneness of Being – of which it is an expression.[4] This universalism is not, however, a reductive dismissal of difference and differentiation. On the contrary, opposition, differentiation, and contradiction are what make up the Oneness of Being: 'the Reality is at once the created Creator and the creating creature. All this is One Essence, at once Unique and Many' (Ibn 'Arabi, 1980: 87). Within the Oneness of Being, the many and the one are thus dialectically fused into a totality driven by constitutive internal oppositions.[5]

4. This he clarifies through the metaphor of a mirror: 'the natural order may … be regarded [at once] as [many] forms in a single mirror or as a single form reflected in many mirrors' (Ibn 'Arabi, 1980: 87).

5. Just how embedded opposition and difference are within this ontology is brought to light in the equally foundational concept of the 'barzakh'. As mentioned in the previous chapter, the barzakh appears in the Qur'an as a transition zone between life and death, and has since been mobilized by a range of scholars to theorize spaces and phenomena defined by liminality, dynamic tension, and the dual qualities of separation and connection (Pettigrew, 2013: 21; Lawrence, 2020; Brower, 2022). For Ibn 'Arabi, the logic of the barzakh is built into the Oneness of Being. In the words of Salman Bashier (2004: 91), 'this one reality, which brings the manifest and the nonmanifest together while keeping them apart, is the reality of the barzakh'. If the concept of the Oneness of Being evinces a non-Eurocentric universalism, its twin concept of the barzakh safeguards against what George Ciccariello-Maher (2017: 10) has called 'the lure of totality'. It prevents, in other words, the false comforts offered by a reductive totalization that erases contradiction and difference.

While the Oneness of Being thus evokes a universalism unshackled from the limits of Eurocentrism, of the kind Amin envisioned, it must be acknowledged that it emerged from a period preceding capital's global expansion. For Amin, this might render it obsolete; the material universalization that has occurred through the global expansion of capital has meant that any universalist outlook that preceded this point was merely potential in form, rather than material (Amin, 1988: 72). The point, for him, is that global capitalist modernity has forged an unprecedented material universalism through the spread of capitalism, while also producing a Eurocentric, false universalist ideology which obfuscates these material transformations.

Today, however, these twin poles of modernity are drifting ever farther apart. For Achille Mbembé (2021), ours is an era of late Eurocentrism, which 'seeks to justify the battening down of Europe on itself, its withdrawal from the world'. Whatever else it may be, this is not an era in which the Eurocentric hubris to which Amin drew attention sits as comfortably over the world system as it did during earlier periods of European dominance. While rife with its own dangers, this moment creates epistemic openings which may not have been apparent at the time of Amin's writing *Eurocentrism*.

If the era of 'late Eurocentrism' is indeed characterized by Europe's withdrawal and battening down, migration and border externalization are central elements of this juncture. For where else can the impulse towards walling off and withdrawal be seen more clearly than in the unprecedented drive towards outsourcing migration and border controls? For this reason, let's briefly dwell upon the growing literature on EU border externalization, drawing from it some analytical strategies for discerning how to draw a wedge between the Eurocentrism of externalization and its analysis. In doing so, we can see more clearly how externalization sits within this juncture, and some renewed epistemic opportunities associated with it.

Distancing strategies

Perhaps unsurprisingly, early scholarly perspectives on EU external migration policies showed little concern with the perspective of those targeted by these policies, and much less their social and historical contexts.[6] The autonomy of

6. Rooted largely in the disciplines of political science and security studies, they highlighted shifting loci of migration management in Europe, first 'upward' from national capitals to the institutions of Brussels (Guiraudon, 2000; Kostakopoulou, 2000; Schain, 2009), and then 'outward' to the domain of foreign relations (Bigo, 2000; Boswell, 2003; Lavenex, 2006; Lavenex and Schimmelfennig, 2009). Driven by an unprecedented securitization of migration in Europe (Huysmans, 2000, 2006), this delocalization led to conceptualizations such as 'policing at a distance' (Bigo and Guild, 2005), 're-scaling' (Samers, 2004), and 'remote control' (Zolberg, 1997, cited in Fitzgerald, 2019). These, in turn, laid the ground for research into the policy instruments and frameworks of externalized EU migration and border governance, such as Mobility Partnerships (Maisenbacher, 2015; Reslow and Vink, 2015; Hampshire, 2016) and readmission agreements (Strik, 2010; Wolff, 2014; Carrera et al., 2016).

migration perspective represented a welcome departure in this regard, as it foregrounded issues of migrant agency and subjectivity (Bojadžijev and Karakayalı, 2010; Tsianos and Karakayalı, 2010; Mezzadra, 2011c; Tazzioli, 2014). In this light, migrants' struggle, resistance, and agency become constitutive drivers of migration and border policy (Ataç, Rygiel and Stierl, 2016; De Genova, 2017a; Karakayali, 2018). In doing so, this perspective successfully forged a distance between the EU border regime and its critical analysis. Applied to externalization, this perspective shows the outsourcing of migration and border controls to be driven in part by the unpredictable turbulence of collective migrant movement (Casas-Cortes, Cobarrubias and Pickles, 2015; cf. Cobarrubias, 2019). This presents a rather less static picture of the EU border regime in West Africa than earlier policy-oriented literature; instead, a fluid and itinerant bordering assemblage emerges, one that operates in reaction to the movements it seeks to control (Casas-Cortes, Cobarrubias and Pickles, 2014: 246).

At the same time, however, certain strands of this work rest upon Eurocentric theoretical premises, as its proponents have acknowledged (Mezzadra, 2011c: 122). One practical manifestation of this has been a foregrounding of Europe-bound movement, at the expense of regional mobility.[7] Aside from ethical or political implications, this forecloses other important questions, such as why so many people do not move elsewhere (Gaibazzi, 2015), or how externalization in West Africa shapes trajectories towards continents other than Europe (Marie and Vammen, 2019). For these reasons, while the autonomy of migration's emphasis upon agency and subjectivity holds potential for the analysis of externalization on the African continent, it needs to be balanced with what Leonie Newhouse (2018) calls a manifold reading of human mobility, which foregoes any presumption of a Europe-bound directionality.

If the analysis of externalization thus requires us to expand our concern beyond those on the move to Europe, the upshot is a need to empirically foreground the African contexts in which EU migration policies today operate (cf. Gaibazzi, Bellagamba and Dunnwald, 2017; Savio Vammen, Cold-Ravnkilde and Lucht, 2022). Ruben Andersson (2014a) offered an early insight into these entanglements, highlighting a paradoxical interplay between Europe-bound migrants and the vast array of actors, institutions, and technologies which take these migrants as their object of intervention. With the analysis of externalization embedded within

7. In a sense, this is to be expected; the autonomy of migration's analytical priority is how collective migrant agency challenges and reorients the control drive of the EU border regime. It makes sense, in this light, to focus on Europe-bound movement, since this is where the border regime's resources are themselves devoted. But as these resources are increasingly focused upon stemming Europe-bound movement at its perceived point of origin (New Keywords Collective, 2015: 75), they increasingly target people irrespective of whether or not they intend to reach Europe, as we will see in Chapters 5 and 7. In such a context, an exclusive focus on Europe-bound movement risks absorbing the assumptions of the border regime itself, which depicts regional mobility as inevitably Europe-bound (Casas-Cortes and Cobarrubias, 2019).

the social context of its manifestation, it becomes possible to discern how this illegalization of migration interacts with gendered and racialized structures of oppression and exploitation (Freedman, 2012; Tyszler, 2019; Gross-Wyrtzen, 2020). Similarly, the role played by local informal economies, political interests, and spatial configurations in shaping the form and even success of external security interests in North and West Africa comes to light (Soto Bermant, 2015; Lecadet, 2017; Zampagni, 2017; Raineri, 2021; Sylla and Cold-Ravnkilde, 2022). So irreducible is the domestic political context and social reality, then, that any investigation that ignores them is likely to miss much of how externalization plays out in practice.

For these reasons, other scholars have foregrounded the role of non-EU state actors, presenting a challenge to the Eurocentric assumptions of the border regime as well as much of the scholarship devoted to it (El Qadim, 2014; Cassarino, 2018; Roman, 2018; Karadag, 2019; Derrider and Pelckmans, 2020). From this perspective, states beyond the external border are not a monolithic landscape defined by their subjection to EU externalization prerogatives. Rather, they are characterized by contestation, ulterior motives, and a non-negligible scope in which these can be achieved (Gross-Wyrtzen and Gazzotti, 2021). Philippe Frowd (2022) similarly challenges the EU's depiction of its security partners, arguing against externalization as a useful analytic for grasping external intervention in West Africa, and highlighting the EU as but one actor within an emergent transnational security assemblage in the region (Frowd, 2018). Such perspectives are particularly relevant in an era of 'migration diplomacy', in which the policy issue of migration control affords an unprecedented degree of diplomatic leverage to non-EU states (Tsourapas, 2017; Adamson and Tsourapas, 2019). As we will see in Chapters 3 and 4, these insights are of significant relevance to Mauritanian state actors, who find ways of aligning the policy objectives of externalization with pre-existing interests and strategies.

That said, while it is reductive and Eurocentric to treat non-EU states as one-dimensional blank slates, placing exclusive emphasis upon their agency in navigating the terms under which externalization is implemented carries its own risks. In particular, one risks losing sight of the fact that this implementation occurs within an asymmetric structure inherited from the colonial era. Scholarship has for this reason increasingly turned to this history, finding in externalization continuities and ruptures with the colonial period (Brambilla, 2014; Ould Moctar, 2020; Stambøl and Jegen, 2022).[8] Indeed, for Martin Lemberg-Pedersen, (2019: 250), it

8. Polly Pallister-Wilkins (2017: 67) lays emphasis upon 'wide temporal periods' in her analysis of the tensions of the Ceuta and Melilla border fences, while Xavier Ferrer-Gallardo (2008) prefaces his analysis of this border with a historical overview commencing with the epochal shift of 1492. Maurice Stierl (2019) also turns to colonial histories of slave resistance in his rebuttal to widespread popular comparisons between people smuggling and the slave trade. A similar grounding is evident in Luna Vives' (2011: 56) 'alternative reading' of the EU's southern border, which discusses how the latter builds upon and sustains much older Spanish and European historical structures of white supremacy.

is precisely through an inquiry into the constitutive effects of colonial encounters upon contemporary policies that the capacity of local actors to either subvert or strategically conform to external dictates is discerned.

From what might be loosely deemed the 'externalization literature', then, four strategies can be drawn for driving a wedge between the Eurocentrism of the process and its scholarly critique. The first consists of highlighting the constitutive and, at times, steering role of migrant agency and subjectivity within the strategies and tactics adopted by the border regime (De Genova, 2017a); the second foregrounds the social relations and structures of the African contexts in which EU migration policies are implemented; the third nuances the tabula rasa picture of non-EU states that is implied within many depictions of externalization as a simple and direct transfer of power from the EU to its outside; and the fourth situates the externalization processes within deeper histories of colonialism, imperialism, and racism. There is a productive tension between these analytical strategies. Absent a focus on situated social relations, for example, the picture of migrant subjectivity that one comes away with can be all too similar to the Eurocentric directional assumptions of the border regime. And while it is folly to discount the role of non-EU state actors in shaping externalization's implementation, it would be equally unwise to disregard the historical structures in which this agency is enacted, as the constraints and opportunities that emerge from these structures often shape the form it takes.

Marrying these productive tensions and working from them, this book puts forth a two-pronged conceptual framework for a non-Eurocentric analysis of the externalization process. The first pole of this framework situates externalization within a deeper history of European conquest and dominance (cf. Danewid, 2017), situating the role of non-EU state actors in this context. This does not diminish the significance of their influence in the externalization process, but rather clarifies some contours of their actions (Chapter 4). The second pole foregrounds the agency and aspirations of migrants, irrespective of whether they are en route to Europe. In the place of this directional assumption, it situates their experience within the social relations that have emerged from this history, and within a social condition prevalent across much of the Global South contexts to which migration controls are externalized.

Externalization, or, the twilight of Eurocentrism

Saying it is necessary to historicize the externalization process is one thing; doing so is another. What, for instance, is a relevant temporal cut-off point when situating a given social process within its deeper history? And how, in doing so, to avoid reducing the entire interim period to a monolithic bloc of time? The following discussion will likely succumb to some of these risks, despite best efforts to overcome them. If the analytical task is one of situating externalization within the history of European dominance, then a logical starting point is the emergence of this dominance, given the foundational role it plays in the material constitution of the modern world system, but also in how we know about this system (Mignolo and Tlostanova, 2006: 208). The following section conceptualizes the emergence of the era of European dominance from the perspective of the Mediterranean, illustrating contemporary rifts in this world systemic arrangement that become apparent from this perspective, and detailing what the vantage point of Mauritania can tell us about these dynamics. We then turn from the colonial era to the period of indirect colonial rule in Africa, highlighting governance strategies of this period which bear significantly upon the form that externalization takes in African contexts such as Mauritania. With externalization situated in relation to these two elements of the history of European dominance, ruptures as well as continuities become apparent. In particular, an unresolved relationship between fascism in Europe and colonialism abroad comes to light, a relationship that is coming to a head in the externalization process.

The historic arc of the Mediterranean centre-periphery boundary

While the southern EU external border did not exist when Samir Amin originally wrote *Eurocentrism*, he accords a significance to the Mediterranean that has only become more prescient in the intervening years. Indeed, the Mediterranean is the cornerstone upon which what he terms the Eurocentric ideological construct rests. Its centrality lies in its dual function of mediating material asymmetries while at the same time providing the Eurocentric edifice with its symbolic foundations: 'The new European culture reconstructs itself around a myth that creates an opposition between an alleged European geographical continuity and the world to the south of the Mediterranean, which forms the new centre/periphery boundary. The whole of Eurocentrism lies in this mythic construct' (Amin, 1988: 11). At one level, then, the Mediterranean is a material boundary that mediates uneven development between an emergent European core and an African continent undergoing progressive peripheralization from the fifteenth century onwards (Rodney, 1972; Amin and Girvan, 1973). The rise of Portuguese naval imperialism in the late fifteenth century was key to this shift, as it gradually subordinated the Mediterranean to the Atlantic (Banaji, 2007: 49). With the sphere of circulation of an embryonic world capitalist system oriented around the Atlantic, the Mediterranean would eventually become a mediating boundary between the centre and the periphery of this new system.

In addition to this material role, however, the Mediterranean also acts as a Eurocentric geographic imaginary that elevates this new centre-periphery economic relationship to the status of a civilizational archetype. For Amin (1988: 93), 'this new North-South split, running through the Mediterranean' is projected backwards into history, endowing it with a potent and ubiquitous ideological sway: 'from the works of apologists for the French colonial conquest to the speeches of Mussolini to the textbooks still in use throughout Europe, this North-South cleavage is presented as permanent, self-evident and inscribed in geography' (Amin, 1988: 93). As well as generating a vast material divide between the centre and the periphery of the world capitalist system, the age of European dominance thus transformed the Mediterranean into a symbolic boundary distinguishing an emergent European identity from its peripheral Other. If Eurocentrism is an ideological superstructure sitting atop the 'economic base' of the world capitalist system, the Mediterranean is thus a space in which these two levels collapse. It is both the mediating zone between the economic core and the periphery and the source of the myth that justifies the very existence of this separation.[9]

If the Mediterranean is both a mediator of uneven economic relations and a geographic imaginary that mystifies this relationship, it is useful to disentangle these two elements of the centre-periphery boundary. As far as the economic dimension is concerned, the intention here is not to claim one way or the other whether the world economy in general continues to exhibit a core-periphery structure.[10] Attention is paid in Chapters 3 and 4, however, to certain features of Mauritania's political economy that reflect the peripheral economic status it acquired during the colonial era, and which endure today. As we will see, this facilitates a more holistic appraisal of the role of Mauritanian state actors in the negotiation of externalization in the present, as it situates their agency within a broader structure of economic dependency. In doing so, the primary aim is to broaden the scope of interest within EU border externalization debates about non-EU state agency, given that such questions of political economy rarely figure within these discussions.

More central to the focus on the Mediterranean centre-periphery boundary here, however, is its immaterial quality as a geographic imaginary. During the Cold War, this North-South divide was grafted onto an equally symbolic East/West divide, with the Mediterranean figuring in the eyes of NATO strategists as their

9. While unique in its framing in *Eurocentrism,* Amin's analysis of the centre-periphery boundary is very much a product of long-standing debate within development studies, and in particular the traditions of dependency theory and world-systems theory (e.g. Prebisch, 1959; Frank, 1969; Wallerstein, 1974; Cardoso, 1977). These traditions shed light on the dynamics of the world capitalist system from the perspective of the formerly colonized world, revealing how colonial economic arrangements survived the era of independence in the form of an economically self-reliant core and an export-dependent periphery (Amin, 1974a). This model has since come under sustained criticism, however. For some, dependency theory has been ill-equipped to analyse the changes that have taken place since the late 1970s in the global economy (Marcussen, 1982; Petras, 1982; Petras and Brill, 1985; Sánchez, 2003). Instead, it confines global economic dynamics to a static power relation, in which agency and fluidity are sacrificed to the pre-determined outcomes of the centre-periphery model (Leys, 1977; Thandika Mkandawire, 2001). Critical border studies scholars have similarly cautioned against too rigid an application of the centre-periphery model, which risks over-simplification of the geographies produced through capital's worldwide expansion (Mezzadra and Neilson, 2013: 74).

10. While Mezzadra and Neilson suggest otherwise, Robert Wade (2018) highlights certain contemporary global trends – patenting, global value chains, financialization – which he argues reproduce the core-periphery structure of the global economy.

'southern flank' (Lutterbeck, 2006: 77). With the fall of the Soviet Union, and the proclaimed end of history that accompanied it, the Eurocentric teleological conceit that 'societies can only progress to the extent that they imitate the West' (Amin, 1988: 107) became virtually axiomatic. As a result, the lure of the Mediterranean geographic imaginary increased further. In the EU context, this was expressed in the Barcelona Process, which birthed an ambitious framework of 'Euro-Mediterranean' partnership agreements that aspired to spread democracy, good governance, and human rights to the societies south of the Mediterranean (Wolff and Pawlak, 2017). In other words, the faux universalism that asserts that 'the only possible future for the world is its progressive Europeanization' (Amin, 1988: 107) was embodied in the association agreements of the Barcelona Process.

That said, there was a tension at play in this process, one that became apparent as the universalizing ambition of the Barcelona Process continued to be asserted even as the drive to 'democratize' Mediterranean partner states increasingly faltered (Youngs, 2002). For Amin, however, there is no contradiction here. 'What is the point', he rhetorically asks (Amin, 1988: 111), 'in exhorting others, "Do as we do", if it is obvious from the start that it is impossible?' The reason is that this false universalism gives cover to, and naturalizes, the ongoing reproduction of the centre-periphery structure. And indeed, as migration concerns figured more centrally within Barcelona Process negotiations (Testas, 2001; Cross, 2011), a centre-periphery geographic imaginary gained increasing traction within EU policy circles. Maribel Casas-Cortes and Sebastian Cobarrubias (2019) trace the immediate roots of this imaginary to a 1998 EU Strategy Paper on Migration and Asylum, which 'literally puts the EU in the center, dictating who should move and who should not move around the world' (Casas-Cortes and Cobarrubias, 2019: 198). From this centre, there emanate four concentric circles, each defined by its relationship to the EU and by its constituent countries' cooperation in the prevention of migration presumed to be oriented towards the EU centre (cf. Balibar, 2009: 199; Casas-Cortes and Cobarrubias, 2019: 198). While the strategy paper was voted down, the geographic imaginary it espoused was nonetheless embraced with the 2004 European Neighbourhood Policy, whose 'Ring of Friends' framework renders the Mediterranean a boundary separating a 'core' of EU Schengen states from their southern partner states (Zaiotti, 2007).

If the Mediterranean centre-periphery boundary has thus acquired further relevance as the EU migration and border regime has developed, one may still justifiably question the point of devoting such close attention to the Mediterranean in a discussion of externalization in Mauritania. Sitting on the Atlantic coast of the Sahara, Mauritania is, after all, farther removed from the dynamics and history of the Mediterranean than, say, Tunisia or even Morocco. And in terms of EU external relations, Mauritania's membership of the bloc of 'Africa Caribbean Pacific' states remains the overarching guarantor of access to EU development financing, trade agreements, and diplomatic channels (Baila, 2016). This is in marked contrast to its North African neighbours that have signed bilateral Euro-Mediterranean association agreements with the EU (cf. Cavatorta and Rivetti, 2014). That said, the

Barcelona Process did encompass Mauritania when it expanded in 2008 in the form of the Union for the Mediterranean (Cardwell, 2011: 229). And as far as migration control is concerned, Mauritania has been considered a Mediterranean partner state since the 1990 Western Mediterranean Forum resulted in the creation of the 5+5 dialogue, an intergovernmental forum bringing together five states north and south of the Mediterranean with the aim of coordinating a variety of migration and border management issues (IOM, 2023).

At a more general level, it is precisely by virtue of being situated at the margins of this Eurocentric geographic imaginary that Mauritania affords unique insight into its peculiarities. As scholars within multiple disciplines have long argued (Tsing, 1994; Enloe, 1996; Mainwaring, 2019), privileged insight into a given social formation – or its geographic imaginary for that matter – can be obtained from its margins. In this regard, Mauritania's historic geopolitical position vis-à-vis the Mediterranean is unique. Indeed, its northern coast bore direct witness to the fifteenth-century tilting of the emergent world system from the Mediterranean to the Atlantic mentioned above, as Portuguese explores saw in the Arguin island off northern Mauritania an enticing access point to trans-Saharan trade (Pettigrew, 2019: 8). Located on the Sahara's Atlantic coast yet tentatively incorporated within the geographic imaginary of the Mediterranean centre-periphery boundary, Mauritania straddles a space through which one world system transitioned to another. From its vantage point, as we will see, glimpses of similar such emergent transitions and transformations in the present can be seen.

In other words, rather than being plagued by stasis and fixity, the picture of the Mediterranean centre-periphery boundary that emerges from the liminal standpoint of Mauritania is rather more fluid and negotiable. As it should be, given the broader evolution of the EU migration and border regime in the twenty-first century. While the Eurocentric universalizing ambition of the 1990s was nonetheless accompanied by a creeping securitization of migration (Huysmans, 2000), this latter trend has increasingly taken the driver's seat in a twenty-first-century context shaped by the war on terror, the 2008 financial collapse, the 2011 Arab Uprisings, and the 2015 'migration crisis'. As a result, while the roots of externalization as an EU migration management strategy lie in the 1990s, it has gained increasing momentum in this twenty-first-century setting. How this trajectory has unfolded within the Mauritanian social context will be detailed in Chapter 4, but in general terms, its most visible and visceral consequence has been the Mediterranean's transformation into the world's deadliest border-zone. Far from being the locus of a myth of benign and enlightened European superiority, the Mediterranean has today become a space in which those seeking refuge in Europe are subject to violent pushbacks, offshore containment, and death and abandonment at sea (Stierl and Dadusc, 2021).

Where does this leave the Eurocentric imaginary that has sat atop the modern world system for much of its existence, and which was invigorated with the process of EU integration and expansion? Put differently, what does it mean when a site of such significance to the Eurocentric ideological construct as the Mediterranean is today, in the words of Nicholas De Genova (2016: 33), a 'macabre death-scape', where migration is governed through death (Squire, 2017)? For scholars and activists working from the concept of the Black Mediterranean, the 'crisis' of migration that has afflicted the EU since 2015 is, at base, a crisis of Eurocentrism (Danewid et al., 2021: 11). This resonates with the observations of Achille Mbembe (2021) discussed above, and his distinction between 'primitive Eurocentrism' and 'late Eurocentrism' – the former characterized by an outward drive towards dominance and conquest, and the latter by withdrawal and battening down. As we will see from the perspective of Mauritania in Chapter 4, one component of this withdrawal has been a gradual contraction of the geographic imaginary of the Mediterranean centre-periphery boundary. While expansive and embracing in the early 1990s, it has receded as externalization proceeds, leaving in its stead a space of abandonment and death at sea, and the more cynical and securitized geographic imaginary of the Sahel on land.

At the same time, such moments also yield opportunities to view histories, contestations, and collective subject positions that are otherwise obscured. It is precisely such standpoints that are highlighted through the lens of the Black Mediterranean (Danewid, 2017; Danewid et al., 2021), and in discussions of the entanglements and connectivity that have long characterized the Mediterranean space (El Qadim and Isleyen, 2023). We will shortly draw from these insights a framework for grasping these epistemic openings. First, however, a further specification of the history in which externalization unfolds is required. For while the above charts a general arc of European dominance over what might broadly be called the colonial era, we have as yet said little about the more specific period of European colonial rule in Africa. This is no less demanding of consideration, as it has conditioned the form that externalization takes in Mauritania in significant ways, as we will see over the next two chapters.

Unearthing the remnants of colonial rule

If situating externalization's Eurocentric narrative within the arc of European dominance sheds light on the evolving form of the Mediterranean centre-periphery boundary, it tells us less of the precise form taken by this dominance at its peak. This is equally crucial in historicizing externalization, as the governance strategies adopted at this time have inflected the postcolonial era with dynamics of violence and exclusion. As Mahmood Mamdani (1996) has shown, the period of indirect rule in Africa was founded upon a legal distinction between ethnicities and races; the latter were subjects of rights, the former bearers of custom. On this basis, ethnic groups were governed by a colonially selected 'chief', the designated bearer of customary authority, who would adjudicate and mete out justice within his community on behalf of the colonial power. Each designated customary authority was envisioned to be autonomous and self-sufficient, the cynical aim being to freeze 'tribes' in an 'authentic' state of tradition in fact bestowed upon them by the coloniser. In the context of West Africa, Mamdani (1996: 86) highlights the French co-optation of Islamic spiritual figures as example, suggesting the 1930 French decision to sanction the opening of Islamic schools (*medresas*) in Mauritania under French aegis as heralding the shift to indirect rule in the region (Pettigrew, 2007; cf. Brenner, 2001).

Combined with the strategy of ethnically delineating customary regimes for ethnic groups, these indirect rule policies amounted to an unprecedented racialization of territorial belonging in Mauritania. As we will see over the course of the next two chapters, the French colonial encounter did not fabricate racial groups in the Senegal River Valley and the Sahara. But it did create a new link between linearly demarcated territories and particular ethno-racial groups. This racialization of territorial belonging undergirded a spate of violence and expulsions in the late 1980s, and it informs who is likely to be deemed deportable by authorities charged with implementing EU migration control prerogatives today.

There is also a more general resonance between the practice of indirect rule in Africa and the manner in which externalization unfolds today. As externalized EU migration governance has expanded over the 2000s, and in particular since the launch of the EU Trust Fund for Africa, it has increasingly been couched in the language of development. As Mark Duffield (2005) has shown, however, such a framing of development – as something that needs to happen in the Global South for the security of the Global North – harks directly back to the period of indirect rule. While temporarily reframed by the geopolitical security stakes of the Cold War, development interventions in the late twentieth and early twenty-first centuries once more rekindled colonial biopolitical security strategies associated with the era of indirect rule. In each instance, this is achieved through fostering self-reproduction strategies on the part of the colonized target group, who are assumed to be capable of autonomous resilience in the absence of the welfare, insurance, and pensions schemes familiar to the Global North. In this light, the development project is more about upholding a colonially endowed global 'life-chance divide' than reducing it (cf. Duffield, 2010). It is for these reasons that Duffield elsewhere (Duffield and Hewitt, 2009) draws connections between the international development project and nineteenth-century liberal colonialism.

Within this broad articulation between the era of indirect rule and contemporary international development practice, the externalization of migration management has a particularly central role to play. Indeed, the containment of unwanted migration flows is a central pillar of what Duffield (2010) refers to as the liberal way of development, given that it is premised upon securing the Global North against those seeking to overcome the global life-chance divide through unauthorized migration. Moreover, as Paolo Gaibazzi (2023) highlights in relation to the IOM's activities in the Gambia, externalization often unfolds through strategies of co-opting and recruiting existing community networks and power structures in a fashion reminiscent of the governance strategies of indirect rule. Eva Magdalena Stambøl and Leonie Jegen (2022), meanwhile, highlight an ongoing overlap between corporate interests and policing practices that was at play in late nineteenth-century French West Africa, which remains observable in the region today.[11]

In drawing these resonances between the colonial era, externalization, and the international development project, there is a risk of reproducing certain Eurocentric practices detailed in the first section in our attempt to undo them. After all, as far as Mauritania is concerned, the colonial and postcolonial periods are but a drop in a vast ocean of regional history (cf. Ould Cheikh, 1985; Bonte, 2008; Pettigrew, 2023). Indeed, almost seven centuries separate the twelfth-century consolidation of the Almoravid empire – which spread the Maliki school of Islamic jurisprudence to Saharan peoples amongst whom it is still exercised today – and the colonial conquest of this region at the turn of the nineteenth century. In placing seemingly disproportionate emphasis upon the colonial encounter in this context, are we not reproducing the Eurocentric habit of endowing Europe with historical priority? It is for such reasons that Camille Lefevbre (2015) lays emphasis upon the continuities and overlaps between precolonial systems of frontier administration and those introduced during the French colonization of Niger.[12] In a similar spirit, the next two chapters trace continuities and ruptures through a timespan far preceding the colonial encounter, detailing the sedimentation of social relations and structures in which the externalization process intervenes.[13] If we are to successfully situate migrant subjectivity within the social relations of its manifestation, as per the aims of the following section, the history of these relations cannot be reduced to a fleeting moment commencing at the turn of the nineteenth century, as the following chapters make clear.

At the same time, however, a core assumption underlying these chapters is that there *is* something foundational at play in the colonial era, one that demands consideration in the analysis of externalization.[14] As suggested above, the colonial legacy of indirect rule conditions the form that externalization takes as it unfolds beyond the European continent. But it also has a discreet relationship to what

11. This comes to light through their analysis of Civipol, the French semi-state agency that implements EU Trust Fund projects in a number of West African contexts, which we will encounter in more depth in Chapter 4.

12. Similarly, in the analysis mentioned above, Paolo Gaibazzi (2023) draws attention to how dynamics and forces originating *beyond* the colonial encounter become enfolded within the IOM's system of 'indirect migration management'. And while distinct in disciplinary terms, Hannah Cross's (2013) concern with the political economy of EU migration policies in West Africa also encompasses regional migration dynamics, precolonial social formations, and colonial-era dispossession processes.

13. In doing so, they also engage with a diverse literature that centres regional social systems, geographies, and histories (Amin, 1974b; Potts, 2010; Hall, 2011; Whitehouse, 2012; Choplin and Lombard, 2014; Rossi, 2015; Vigneswaran and Quirk, 2015).

14. This is not least for epistemological reasons, given the nature of most archival sources available and the categories of knowledge they employ (Ould Cheikh, 2014b; Anderson, 2021).

drives externalization in the first place. To bring relationship to light, however, we must briefly consider the role externalization plays vis-à-vis the spectre of fascism in Europe.

Externalizing the return of the repressed

The drive to keep the far-right at the electoral margins has long been a crucial impetus to externalization. The 2002 European Council meeting in Seville, for instance, was a key moment in the integration of migration management into EU external relations (Gabrielli, 2016: 9), as its conclusions for the first time pledged all future association agreements with non-EU states to contain clauses on compulsory readmission and joint management of migration flows (Lavenex, 2006: 342). Impelling this incipient externalization within the Seville Council conclusions were significant far-right electoral gains in France and the Netherlands earlier that year (Peers, 2006: 108; Schain, 2009: 103).[15] Now commonplace, this strategy of offsetting fascist sentiment by channelling it into EU external relations has led Gregory Feldman (2011b: 25) to conceptualize the EU border regime in terms of a dynamic tension between 'neo-liberals and neo-nationalists' (cf. Ferrer-Gallardo and Gabrielli, 2022). And as we will see in Chapter 4, it remains a justification of externalization in the eyes of EU officials based in Mauritania.

What does this strategy have to do with colonial rule in Africa? In his discussion of a massacre of migrants in June 2022 at the Spanish-Moroccan border of Melilla, Ben Ehrenreich (2023) suggests that 'fascism in Europe cannot be separated from violence at its colonial frontiers'. This observation emerges from a reflection on the history of the Melilla border. Long before it was fortified with EU-funded security fences and infrared cameras, Melilla was the site of an uprising that was brutally put down by a Spanish colonial army, one whose cadres went on to stage the coup that launched the Spanish Civil War and Franco's subsequent dictatorship. This is but one concrete example of Aimée Césaire's (2000: 36) famous observation that fascism in Europe is colonialism turned inwards. In this light, externalization is a bulwark against internal colonialism, against the possibility that the colonial hierarchies and violence whose residues it sustains may eventually manifest in Europe. As such, it exhibits a key feature of the system Kwame Nkrumah (1965: xii) described as neo-colonialism, namely its habit of 'exporting the social conflicts of capitalist countries'.

Both Nkrumah and Césaire were acutely aware of the impossibility of such a state of affairs existing in perpetuity. As an inherently unstable – not to mention inhumane – arrangement, it is bound to be undone by its own contradictions. As noted in the previous section, the Eurocentric geographic imaginary of the Mediterranean centre-periphery boundary has contracted as externalization

15. In France, FN candidate Jean-Marie Le Pen for the first time made it to the run-off stage in presidential elections against Jacque Chirac, while in the Netherlands the anti-Islam LPF party entered into a short-lived coalition government after the elections.

proceeds, transforming it from the source of the myth of European civilizational superiority into a space of abandonment and death. This Eurocentric self-image is equally eroded in externalization's drive to outsource the threat of fascism beyond Europe's borders. As will be further argued in Chapter 4, the practical effect of this strategy is a growing coalescence between externalization's material consequences and fascist fantasy. Amin, for his part, was aware of the potential for such affinities to develop, observing that 'Nazism, far from being an aberration, always remains a latent possibility, for it is only the extreme formulation of the theses of Eurocentrism' (Amin, 1988: 114).

Our historicization of externalization thus poses with particular acuity the question of a genuine, non-Eurocentric universalism, given the increasingly perverse forms taken by its Eurocentric variant.[16] Can the contemporary juncture of late Eurocentrism at last cede ground to a more encompassing universalism? And what does the perspective of those on the receiving end of border violence indicate about this potential? In the following section, we turn from history to social relations, drawing some tools for conceptualizing in general terms the relations and structures in which migrant subjectivity is forged within Global South contexts such as Mauritania.

Social relations and subject positions in Europe's periphery

The above has hopefully illustrated how externalization can be situated in a deeper history of European dominance and conquest without reifying or succumbing to the Eurocentric conceits in which this history is often shrouded. But any analysis of externalization that stops here would remain Eurocentric in another respect, for it has as yet paid little attention to the people themselves targeted by externalization policies. Indeed, for Meera Sabaratnam (2017), the tendency to analytically sideline the people targeted by external interventions is a key vector for Eurocentrism. Crucially, in her reading, Mark Duffield's work also falls prey to such Eurocentric conceits. In tracing the biopolitical population management strategies of Western interventions from the colonial era to the present, she suggests, he leaves little scope for agency, contestation, or indeed mere existence, on the part of those targeted for intervention. For Sabaratnam, this assumption does not just represent a set of epistemic priorities that remain uncomfortably aligned with those of the interveners themselves; it also obscures central elements of empirical reality. Integrating the viewpoint of those targeted by external intervention results, in other words, in a more comprehensive perspective on the 'totality' (Sabaratnam, 2017: 51). In what follows, we draw out some conceptual tools for integrating the perspective of those illegalized through externalization into its analysis, situating this perspective within the social relations and structures

16. It is no coincidence that in the same passage mentioned above, Césaire (2000: 37) goes on to disparage the 'pseudo-humanism' that has long given cover to colonial violence.

of many Global South contexts to which migration controls are externalized. With this perspective in the foreground, we can discern not only how externalization interacts with pre-existing relations and structures; the consequent viewpoint on the totality also offers means of grasping epistemic opportunities within the juncture of European decline.

Illegality and living labour

To place the perspective of those targeted by externalization at the heart of this inquiry is to ask how and why particular swathes of people come to be viewed as in need of containment outside Europe. As we have seen, within the border regime's Eurocentric imaginary, the world south of the Mediterranean is defined by its status as a potential source of Europe-bound migration. The cartographic instruments of externalization – such as the ICMPD's i-Map – are explicit in this regard, as Sebastian Cobarrubias (2019) has shown. Europe-bound movement is presented as a pre-existing 'fact on the ground', one that strives to contain this movement through 'restrictive practices and discourses for governing migration at a distance' (Cobarrubias, 2019: 777). The enactment of these practices and discourses, in turn, produces more of the 'irregular migration' that they claim to prevent, as Ruben Andersson's (2014a) ethnography of the illegality industry shows.[17]

If the externalization process reduces the world south of the Mediterranean to a potential source point of Europe-bound migration, this means it is not just those actively on the move to Europe whom it views as in need of containment. In a context of border control policies informed by strategies of 'pre-frontier detection' (New Keywords Collective, 2015: 75), ordinary individuals are inevitably recast as 'potential candidates for irregular migration' (cf. Migreurop, 2010), irrespective of their actual itinerary or subjective desires. Indeed, as Andersson (2014a: 116) observes, police forces in Mauritania and Morocco often expand the targeting scope of their operations along racial lines, for the sake of keeping up appearances in a context of EU pressure to police 'irregular migration'. If we are to centre the perspective of those targeted in such operations, it is thus essential to shed the assumption that migrants are inherently and inevitably *en route* to Europe. As we will see in Chapter 6, once this assumption is discarded, the relationship of migrants who are elsewhere illegalized by the border regime to regional histories and social dynamics is brought into relief. In this manner, the 'colonial erasure' that Cobarrubias (2019: 778) argues to be built into externalization's geographic imaginary is overcome, and clues offered as to who is likely to be rendered 'illegal' in the first place.

17. This production of 'migrant illegality' occurs not just through the fact that the closure of legal avenues to migrate generates a market for illegal ones; 'illegality' also becomes embodied as a subjective condition, as those seeking to evade capture evince behaviours associated by authorities with the 'irregular migration' they are tasked with policing (Andersson, 2014a: 120).

To read these signs, however, we must further problematize this condition of illegality, which is, after all, a highly contingent one.[18] Are there more deep-seated subject positions underpinning this surface condition? Sandro Mezzadra's (2011c) 'gaze of autonomy' offers insight in this regard, as it locates migrant illegality 'within a wider analytical framework that examines the transformations of contemporary capitalism from the point of view of living labour and its subjectivity' (Mezzadra, 2011c: 137). Illegality is here situated within a more general subject position: that of living labour. Memorably juxtaposed by Marx (1906: 256) with capital, which is dead labour that feeds upon the living, the standpoint of living labour brings out a crucial contradiction that inheres in labour as a commodity; humans are at the same time an objective source of surplus value to capital and endowed with the subjective capacity to challenge their exploitation. Labour is a commodity, albeit one endowed with human agency.[19] Agency and subjectivity are, for this reason, of crucial analytical and political importance; they offer unique insight into social relations and structures, without falling into deterministic readings of them. For these reasons, foregrounding the agency and subjectivity of living labour allows us to forge a distance between the Eurocentric assumptions of the border regime, and instead pay attention to what emerges when it periodically infuses illegality into social structures in Global South contexts such as Mauritania.

To do so, however, we must also consider such Global South social structures in general terms. This, in turn, means moving beyond the European 'location' that Mezzadra (2011c: 122) admits his 'gaze of autonomy' to be located. As he elsewhere argues, any global picture of migration and capitalist development must account for the heterogeneity of forms that labour relations take across the Global South (Mezzadra, 2011b: 165). In other words, just as it is necessary to expand the sphere of concern beyond Europe-bound movement when centring the perspective of those at the receiving end of externalization, so too must we drop any assumption that formal 'free' wage labour constitutes a gold standard or

18. This social contingency is well detailed in the old and vast literature on the topic of migrant illegality (cf. Calavita and Honig, 2005), which has variously drawn attention to: ambivalent spaces of 'liminal legality' (Menjívar, 2006; Chacon, 2015; Feldman, 2020); the continuum of 'probationary citizenship' in which illegality sits (Chauvin and Garcés-Mascareñas, 2012); forms of urban civic and community participation that foment an 'unauthorized yet recognized' political subject (Sassen, 2005: 85); the institutional means by which deportation becomes deferred (Castañeda, 2010); and the grey space of 'entanglements of formal and informal realms' in which illegality becomes socially enmeshed (Reeves, 2013: 511). It is for these reasons that, in a seminal reflection on the condition of 'deportability', Nicholas De Genova (2002: 422) suggests that 'to conduct research on undocumented migrants as such – conceptualized in isolation – is … to perpetrate a rather egregious kind of epistemic violence on the social reality of everyday life for those migrants'.

19. This is 'the peculiar nature of the commodity sold', highlighted by Marx (1906: 259), which 'implies a limit to its consumption by the purchaser'.

benchmark for defining capitalism (Mezzadri, 2022). A final step in overcoming Eurocentrism in the study of externalization is thus to consider the form that capitalist development takes in Global South contexts such as Mauritania.

Postcolonial capitalism and racialized surplus populations

As noted above, illegalized migrants in Mauritania generally get by on the urban margins, floating between unemployment and an array of jobs collectively – and somewhat arbitrarily – deemed 'informal'. While precise numbers are by definition difficult to pin down, official sources widely acknowledge that the vast majority of migrants in Mauritania work informally (IOM, 2009: 16; Islamic Republic of Mauritania, 2010: 29). Moreover, this informal sector was in 2017 estimated to make up 91.5 per cent of non-agricultural private sector activity (Islamic Republic of Mauritania, 2017: 6). This is in keeping with the regional average in West Africa, where the figure stands at 87 per cent (Meagher, 2022: 1209). It is for these reasons that Horman Chitonge (2017) argues that 'restricting capitalist exploitative relations to wage labour dynamics yields an incomplete story of how capitalism has unfolded [in Africa]'. Clearly, the informal economy in this context is not a residual or outlier feature to a capitalism otherwise defined by formal wage-labour relations. It is itself of constitutive of capitalism in West Africa. How to further conceptualize the informal economy in this light? And how does externalization's production of migrant illegality sit in relation to it?

An indirect answer to these questions can be found in the work of Kalyan Sanyal (2007), for whom capitalist development in postcolonial contexts is marked by a dynamic contradiction between what he calls the 'accumulation economy' and the 'needs economy'.[20] The former is the growth pole of the postcolonial national economy, and it typically only employs a small segment of a given national population, who thereby constitute the formal working class. The rest are rendered surplus – 'castaways of development', consigned to the 'wasteland' of the needs economy (Sanyal, 2007: 46). Having been uprooted through forms of dispossession such as those that will be detailed in Chapter 6, they are forced to get by in the needs economy through whatever survival strategies they can (Li, 2017: 1249). Crucially, for Sanyal, this needs economy is not capitalist in form; it is the domain of 'non-capital', produced by capitalism as a necessary condition of its ongoing reproduction. The informal sector, then, is none other than this non-capital needs economy, the 'holding zone' to which those rendered surplus to capital are relegated (Bhattacharya, Kesar and Mehra, 2023: 150).

Like Duffield, Sanyal considers international development interventions in this postcolonial context, also seeing in them a biopolitical technology aimed at

20. This is in contrast to the experience of capitalist development in the Global North, whose foundational contradiction is between wage labour and capital, between the working class and the capitalist class. In postcolonial contexts, on the other hand, this labour-capital contradiction is only to be found in the accumulation economy.

fostering resilience and self-reliance amongst those who have been cast out of the capitalist development process (Sanyal, 2007: 126). For Duffield, however, the existence of vast swathes of 'non-insured' life across the globe is taken for granted; it is a background feature of the landscape in which international development intervention takes place. Sanyal, on the other hand, brings this feature into relief, illustrating the specificities of the postcolonial economy that excludes vast swathes of people from its operations.[21] If Duffield's work tells us how externalization upholds a colonially endowed form of development practice that strives to contain the dispossessed in the peripheries, Sanyal illustrates the structural role that their dispossession plays in these peripheries. In a word, the 'candidates for irregular migration' whom the externalization process targets for containment are the product of postcolonial capitalist development, and its creation of vast surplus populations as a condition of its existence.[22]

That this group of people is viewed as a threat in need of containment is more or less openly acknowledged within EU policy documents. The strategic orientation document of the EU Trust Fund, for example, observes with more than a hint of foreboding that '330 million young people will enter the labour market in the next 15 years in Sub-Saharan Africa. The challenge of their socio-economic integration is immense' (European Union, 2016: 4). It must be noted that the externalization process is no mere passive bystander to these dynamics of postcolonial capitalism. As Ruben Andersson (2014a) has argued, the EU border regime requires migrant illegality in order to justify its own existence and produces more of it to this end. If this is so, the border regime achieves this through a nefarious partnership with the wasteful dynamics of a global economic system that nourishes an ever-expanding informal economy as one of its core structural features.

Of course, in the landscape of postcolonial capitalism, the forces determining both who is dispossessed in the first place and the survival strategies they subsequently adopt are deeply gendered and racialized (Bhattacharyya, 2018;

21. As Geoffrey Aung (2018: 198) puts it in his application of Sanyal's framework to the Dawei special economic zone in Myanmar, 'these interventions enable basic subsistence for people dispossessed, who are increasingly likely to be excluded from the formal reproduction of capital'.

22. Scholars have already drawn attention to how the EU border regime's illegalization of migrant labour creates a surplus population from which employers in the Spanish and Italian agricultural sectors can draw (Cross, 2013; Pradella and Taghdisi Rad, 2017; Pradella and Cillo, 2021). Externalization is, in this reading, a means of ensuring a cheap reserve army of labour to capitalists in Southern Europe. While this book does not deny these aspects of how the EU border regime serves the interests of capital in Europe, it is more concerned with what happens on the other side of the external borderline, in the urban informal economy of Mauritania. In this context, distinct claims that the sheer scale of the contemporary surplus population exceeds the reserve army function typically attributed to it by Marxist scholarship appear equally tenable (Li, 2017: 1251; Bhattacharya, Kesar and Mehra, 2023: 149).

Mezzadri, 2022). In the Mauritanian context, these racializing forces hark back to the colonial territorial and governance strategies detailed above. As we will see over the course of the following chapters, the racialization of territorial belonging that occurred in the colonial era (Chapter 3) converges with the dynamics of the informal economy to create a racialized surplus population, one that provides ample human material for the border regime to illegalize (Chapters 6 and 7).

In this light, we can now offer a general response to the question posed earlier about how and why particular swathes of people come to be viewed as in need of containment south of the Mediterranean centre-periphery boundary. Put briefly, the answer lies in the convergence between postcolonial capitalism's reproduction of a structural wasteland of surplus populations and the ongoing legacy of the colonial racialization of territorial belonging. As we will see, by virtue of occupying this structural position, people are subject to a range of bordering and dispersal tactics at the level of the urban informal economy which bear a striking resemblance to externalization measures. This similarity is no coincidence; it is rooted in the fact that the condition of illegality is drawn from the more general one of the racialized surplus population.

Conclusion

As a prevailing EU strategy for governing migration, border externalization has become rather more than the sum of its parts. Once policies aimed at preventing people from migrating at source are put into place, they necessarily come into contact with pre-given social relations and structures. At the same time, a colonial history of European conquest and expansion reverberates in the contemporary drive to outsource migration and border policies. And in the implementation of these policies, the territorial and governance legacies of this colonial history are often revived. It should be little surprise that EU external border policies are themselves underpinned by a Eurocentric conception of Europe's outside, and that they therefore have little regard for these aspects of their implementation. But critical scholarship ought to repudiate such assumptions if it hopes to take stock of the encounter between externalization and the landscapes of its implementation with any kind of rigour.

This chapter has proposed doing so by situating both the externalization process and the condition of migrant illegality within more general structures – that of the Eurocentric geographic imaginary of the Mediterranean in the case of externalization, and the condition of the racialized surplus population in the case of the illegality it produces. From this perspective, externalization's relationship to a broader juncture of European decline and withdrawal becomes apparent. This can be seen in the Mediterranean's transformation into the world's deadliest border-zone, and in its contraction as a Eurocentric geographic imaginary underpinning EU external relations. At the same time, the more securitized imaginary of the Sahel has expanded, with Mauritania figuring more prominently in this capacity in recent years than as a Mediterranean partner state. Amongst the key internal drivers of externalization, moreover, is an attempt to outsource fascist sentiment beyond Europe. Insofar as it represents an attempt to prevent Europe's colonial history manifesting internally, this internal driver of externalization similarly signals a flailing in the dominance by which earlier periods of European outward expansion were defined.

All of this casts a compelling light on the contemporary potential for a non-Eurocentric universalism, of the kind Amin called for when he wrote *Eurocentrism*. At that time, the Eurocentric universalism that asserts a natural affinity between capitalism's expansion and European civilizational superiority was rather more self-assured than it is at the present moment. In this respect, foregrounding the agency and subjectivity of those illegalized through externalization not only offers insight into the latter's reliance upon the machinations of postcolonial capitalism to furnish it with the illegality it requires. Such a standpoint also offers insight into the epistemic opportunities that lie in the contemporary juncture. It is in this capacity that Ibn Arabi's concept of the Oneness of Being makes a fleeting re-appearance at the end of Chapter 7, in the form of a subtle but crucial aspect of the subjective experience of those constructed as a threat by the Eurocentric imaginary of the border regime. By virtue of being attuned to the social totality derived from the Oneness of Being, this aspect of subjectivity points to a universalism above and beyond the Eurocentric offshoot, whose demise is heralded in the externalization process.

Chapter 3

THE PROXIMITY OF THE PAST IN MAURITANIA

The Raft of the Medusa (Figure 2) is a stark and haunting image. Painted by Théodore Géricault over 1818 and 1819, it represents a moment in which the few surviving occupants of a makeshift raft believe their rescue to be imminent, after spending thirteen days cast adrift off the coasts of modern-day Mauritania. They found themselves in this situation following the shipwreck of their frigate, the Medusa, during a mission to Saint Louis to re-establish French colonial rule in Senegal. Sitting atop the mass of bodies, seemingly the most animate of them all, is a Black African man waving a piece of tattered fabric to the *Argus* rescue ship on the horizon. In this regard, this particular rescue image appears an aesthetic inversion of those Ruben Andersson (2014a: 150) suggests to be 'at the heart of the spectacle of the border' which manifested on these same West African coastal waters in 2006. In its latter guise, the rescue image typically involves a 'one-dimensional' depiction of 'a bare, naked, drowning life' (Andersson, 2014a: 151). This has the effect of creating a 'virtuous circle … where the task of patrolling, caring for, and informing on clandestine migration blur into one another' (Andersson, 2014a: 142). Such one-dimensional depictions are epitomized in the award-winning photo of a man staring up in terror from the water as he clutches onto the boat from which the shot is taken, utterly at the mercy of the waves and his Guardia Civil saviours. In *The Raft of the Medusa*, on the other hand, the Black African man sits at the apex of the image with his muscular back to the viewer, and in this position appears an agent of not just his own rescue, but also that of the multi-racial pile of destitute survivors beneath him.

This purposive positioning is likely a reflection of Géricault's abolitionist sympathies (Harris, 2006: 602). And his decision to capture this moment in the first place speaks to an anti-imperial sentiment; survivors claimed that the catastrophe was caused by the incompetence of the captain, a royal sympathizer appointed for his political stance rather than naval experience. The painting is, however, more ambivalent towards the colonial context of the shipwreck. Julien-Désiré Schmaltz, the man appointed governor of the colony of Senegal following its re-acquisition

Portions of this chapter appear in an article in the journal *Anthropologie et développement*, in a special issue on EU external border policies in West Africa (Ould Moctar, 2020).

Figure 2 *The Raft of the Medusa,* Théodore Géricault, 1819. From Wikimedia Commons: https://commons.wikimedia.org/wiki/Category:The_Raft_of_the_Medusa.

by France, was also aboard the Medusa at the time of its wreckage. But being of relatively high standing, he was among the first to be offered a lifeboat and was therefore spared the ordeal experienced by the subjects of the painting. Instead, he continued on his colonial mission to Senegal relatively unperturbed, in the process drifting beyond the realms of Géricault's attention. The colonial enterprise is, in this light, a discreet canvass onto which explicit antagonisms between royalists and liberals, abolitionists and slaveowners, are splashed. In much the same way, the unnamed subtext of the contemporary rescue image is the militarization of migration control, which remains subtly buried beneath the binary imperative of care and control it so urgently invokes (Andersson, 2014a: 139).

In centring the subtexts of each of these historic moments in turn, this chapter and the following one together bring out parallels, resonances, and divergences between them. In other words, this chapter is the first step in our historicization of the EU border regime in Mauritania. As argued in the previous chapter, one way of evading the Eurocentrism of the border regime in its analysis is to delve into the deeper histories into which it intervenes. From this perspective, the export of territorial and population governance strategies from Europe to West Africa far precedes the twenty-first century. Indeed, the period of indirect colonial rule in Mauritania introduced a new, linear model of territorial distinction to the region, which created an equally novel link between racial group and territory. This, in turn, laid a foundation for who would be considered to belong within the Mauritanian nation-state in the present. This colonial export of the border form of spatial organization was itself an outgrowth of the reorientation of the world system's centre of gravity from the Mediterranean to the Atlantic, and the consequent creation of a spatial and ideological centre-periphery boundary running through the Mediterranean (Amin, 1988). As we will see in the next chapter, both this deep-seated history of the Mediterranean and the more recent one of colonial rule each bear upon the externalization process in the present.

This chapter develops these points by tracing the evolution of three interconnected dimensions of territorial organization throughout the precolonial, colonial, and postcolonial periods. In broad terms, these three territorial dimensions are spatial boundaries, human mobility, and group identity and belonging. Each of the three time periods around which the chapter is structured is further divided into three sections, each devoted to an analysis of one of these territorial dimensions. The three dimensions of territorial organization are thus, in a sense, an operationalization of the deeper historical processes into which the EU border regime intervenes. In this light, the transition from the precolonial to the colonial era is marked by a qualitative shift within each of these domains of territorial organization. Having been inaugurated by the colonial externalization of the border, this shift continues to condition their development in the postcolonial context. This narrative culminates in 'the events' of 1989–91, a period of racialized violence and mass deportations that escalated into a border war between Mauritania and Senegal. 'The events' further reified the colonial shift in territorial organization, insofar as the Senegal River became hardened into a racialized line of territorial division, beyond which those deemed racially 'outside' were deported.

Boundaries, mobility, and group identity in the precolonial period

The social structure of precolonial 'Arab Moor', or *biḍân*, society hinges upon two elite pillars: a warrior class originating from a nomadic group in the Arabian Peninsula called the *banî ḥassân*, and a clerical class, named the *zwâya*, of primarily *Sanhâja* Berber origin. Beneath these two dominant groups sits a diverse social stratum, comprising tributary pastoralists (*aznâga*), artisanal workers (*m'aalmîn*), and poets, musicians, and storytellers (*îggâwen*). These otherwise diverse spheres of economic activity have in common the meagre degree of social prestige attached to them (Ould Cheikh, 1985). On the bottom rung of this social order lies a historically enslaved class, named the *harâtîn*.[1] Also referred to as 'black Moors', the *ḥaraṭîn* are often distinguished from their 'white' – *biḍân* – counterparts (Stewart, 1973; cf. Hall, 2011). Farther south, on the Senegal River Valley, kingdoms and

1. Defined by their historic status of 'freed slaves', many within the *ḥaraṭîn* community today experience myriad social disadvantages and forms of marginalization associated with the legacies of slavery, not to mention its persistence within more remote parts of the country.

social groups differed from their northern counterparts in terms of livelihood but rather resemble them in social stratification terms. At the mouth of the Senegal River, within the Wolof-speaking Waalo Kingdom (Barry, 1972), the social system consisted of a free nobility, an artisanal stratum, and an enslaved class. Meanwhile, the Futa Toro kingdom on the middle Senegal River Valley also contained a clerical elite – the *toorodBe* – artisanal classes, and slaves. Crucially, unlike the spatial demarcations that would characterize the colonial period, the territories of the precolonial Futa Toro kingdom encompassed both the north and south banks of the Senegal River. Let's delve further into these precolonial territorial boundaries, examining how they relate to social relations and structures in the spaces of what would eventually become the Mauritanian state.

The itinerancy of territorial boundaries

In both the Sahara and the Senegal River Valley, precolonial territorial boundaries are characterized by fluidity and malleability. To discern the nature of territorial control and boundaries in the Saharan context, however, we must first consider two other modes of social organization that operated at distinct scales in the region. The first of these, at the lower scale, is the *qabîla*.[2] Defined by Abdel Wedoud Ould Cheikh (1995: 3) as 'politico-genealogical and territorial units' which are traced back to a real or imagined common ancestor, the *qabîla* still acts today as a common reference point in Mauritania, if not in a geographic then at least in a social sense. But in the precolonial period, the intrinsic relationship between *qabā'il* and particular expanses of land meant that there was a degree of localized and decentralized territorial control exercised by different *qabā'il*. This territorial control was not fixed spatially, however, as the seasonal dictates of a pastoral nomadic lifestyle meant that the territorial contours of the *qabîla* would shift over the course of a given year.

The *qabîla* can thus be seen as an example of what Achille Mbembé (2000: 263) calls precolonial 'itinerant territoriality', in that 'political entities were not delimited by boundaries in the classical sense of the term, but rather by an imbrication of multiple spaces constantly joined, disjoined, and recombined'. In the Mauritanian context, the itinerancy of these social and territorial units did not preclude the possibility of territorial infractions; in the absence of neighbouring alliances, these could incur raids on livestock holdings and other such economic penalties. This raises issues of protection, governance, and territorial enforcement, which in turn speaks to the next scale of social organization in the precolonial Saharan context, the Emirate system.

Following the victory of *ḥassâni* warrior *qabā'il* over clerical *zwâya* groupings that brought an end to the Char Bouba war of 1644–77, the lands stretching

2. Singular. The plural is *qabā'il*. It is more commonly translated as 'tribe.' Much like the word 'Moor', it is common to speak of *tribus* in Mauritania, but given that the word 'tribe' carries less precision and more baggage, I use the terms *qabîla* and *qabā'il* for analytical clarity.

eastwards from the Atlantic Ocean to Timbuktu and northwards from the Senegal River Valley into the Sahara gradually came to be governed by four Emirates. Immediately north of the western banks of the Senegal River were the Emirates of Trarza and Brakna, named after the leaders who emerged victorious from the Char Bouba war. Farther north was the Emirate of Adrar, which was consolidated in the middle of the eighteenth century (Bonte, 2008), and in the middle of the region, that of Tagant. This is not to say that there was a monopoly on territorial consolidation across this vast region, and much less that there were discernible lines in place demarcating the end of one territory and the beginning of another. Indeed, being separated by zones that could be vast in their geographical expanse, the geographical parameters of the Emirates remain only approximately understood today (Ould Beidy, 2009: 27). Mohamed Mukhtar Ould As-Sa'ad (1989) underlines this territorial fluidity in his history of the Trarza Emirate in the eighteenth and nineteenth centuries. In spite of a wealth of research on the Emirate's history, he notes a virtual absence of any juridical definition of its territory (Ould As-Sa'ad, 1989: 57). This suggests the social distribution of precolonial authority to have been more rigid and discernible than its territorial distribution. The former was defined by the preordained hierarchies of caste and *qabîla*, in contrast to the fluidity and malleability of the latter. There was, however, a territorial distribution of authority, even including a political capital of sorts in the form of the Emir's encampment, called the *maḥṣar*. But this spatial distribution of power lacked the clearly demarcated boundaries that would come to define colonial and postcolonial modes of territoriality.

Precolonial territorial boundaries on the Senegal River Valley were equally fluid and malleable. Measuring 250 miles from west to east but only fifteen from north to south, the spatial dispersal of the Futa Toro state's political power in the nineteenth century would appear to be informed by the relationship of the land to the Senegal River, and more specifically to its rich floodplain. David Robinson (1975: 185) makes passing reference to Futa Toro's boundaries, in his observation that its territories fade into the desert in the north and into the wilderness of Ferlo in the south. Oumar Kane (1974: 237) draws a more precise picture in his observation that 'the authority of the *satigi*[3] of the 18th century is confined to the zone delimited by the wadi of Niorde, the mountains of Wa-Wa and Armetou, the wadi of Ketchi and Lake Rkiz'.[4] This cartographical sketch adds further weight to Robinson's conflation of territorial boundaries with those delimited by the natural environment. Equally in the case of Waalo, Boubacar Barry (1972) mentions such 'natural' landmarks as ravines, baobab trees, riverbanks, and wells when sketching the territorial interface between Waalo and its neighbouring kingdoms. In the precolonial Senegal River Valley, territories were thus also distinguished by zones, often carved out of the natural environment, rather than lines.

3. The *satigi* was the name of the aristocratic classes of the Futa Toro region from the fifteenth century to the time of their deposal and replacement with the *torodBe* in the late eighteenth century.

4. I have translated to English any direct quotations from French language sources.

For these reasons, the boundaries of precolonial authority were just as malleable around the Senegal River Valley as those of their northern counterparts. Reading these same scholars, it becomes clear that loyalties within the peripheral zones of the Futa Toro state and the Waalo Kingdom had a tendency to shift in accordance with the tributary demands made upon populations by ruling elites, as well as the latter's attempts to expand their domain of influence (Kane, 1974; Ould Saad, 2004). Boubacar Barry (1972), for example, illustrates how the boundaries of the Waalo kingdom receded in the latter half of the eighteenth century following a military campaign led by the leader of Futa Toro, resulting in the frontier villages of N'Dierba, Bokhal, Gaya, and Réfo being subsumed within the territory of Futa. The geographical areas that lay between different regional governing entities were thus just as ambivalent and amorphous in their fealties as those that separated the *biḍân* Emirates farther north, meaning that territorial boundaries also had a capacity to expand and contract in accordance with the reach of governing power.

These historians of the Senegal River Valley do not, however, lay analytical emphasis on territorial frontiers in the same manner as As-Sa'ad in his analysis of the Trarza Emirate. This is likely because the sedentary lifestyle of the majority of the region's inhabitants does not problematize the territorial boundary in the same manner as the nomadism that characterized precolonial Saharan livelihoods. Territorial boundaries are, in other words, inseparable from human mobility, the second dimension of precolonial territorial organization under consideration.

Human mobility as social foundation

It should come as no surprise that human mobility was a primordial social element of the nomadic communities of the Sahara. As noted above, cumulative migrations from the seventh century of nomadic *banî ḥassân* from the Arabian Peninsula to the Western Sahara Desert contributed to the constitution of a *biḍân* warrior-clerical elite, whose remnants and internal power dynamics continue to shape political contestations today (Pettigrew, 2021). But on the Senegal River Valley too, human mobility was a foundational component of social structure and relations. The Fulani demographic base of the Futa Toro kingdom, for instance, consisted primarily of inhabitants of a wide range of surrounding regions, who came to settle on the middle Senegal River Valley over the course of the fourteenth and fifteenth centuries. Migrations were equally influential in shifting the internal demographic dynamics of Futa Toro over the course of its history, as drought and war spurred a movement of peoples within Futa Toro in the early seventeenth century (Robinson, 1975: 193). Vanquished political factions also fled Futa Toro to settle in neighbouring areas, some temporarily to plot their return, others becoming permanent residents within their host communities (Curtin, 1971: 19).

In the Sahara, the nomadism of the majority of the desert's inhabitants was intimately connected to the fluidity of territorial boundaries and control outlined above. Indeed, both Ould Beidy and As-Sa'ad attribute the lack of fixity of the Emirates' boundaries to the spatial mobility by which those who fell under their authority gained their livelihood. Furthermore, the primary source of economic value within such nomadic communities was not land, but herds (Bennoune, 1978: 33); intensive livestock raising formed the subsistence base of nomadic tributary groups and was a key source of revenue extraction for *ḥassâni* and *zwâya* elites. For this reason, while ownership of livestock may have been clearly defined within this precolonial context, a similar emphasis on demarcating land was absent. The link between modes of production and particular configurations and representations of space, which leads Mezzadra and Neilson (2013: 30) to speak of a 'primitive accumulation of modern cartography', is thus clearly at play here. The relationship of this nomadic human mobility to wealth accumulation is not clear-cut, however. As Riccardo Ciavolella (2011) demonstrates, the precolonial mobility of Fulani populations gradually became static as a result of marginalization associated with ecological and political crisis. On the other hand, the diffuse and mobile positioning of certain *zwâya* actors within the trans-Saharan caravan trade would allow them to take full advantage of the emergence of capitalism in the region (Bensaâd, 2011).

While human mobility was not regulated in the bureaucratic manner that would come to characterize the colonial era, it was nonetheless 'channelled', to use Vigneswaran and Quirk's (2015) phrasing. This channelling of mobility was a further factor in the malleable and decentralized form of precolonial territorial control in the Sahara outlined above. In the words of Mahfoudh Bennoune (1978: 34): 'merchants paid dues at times to powerful and strategically located nomadic communities for the right to "pass back and forth" with their caravans.' Channelling mobility was thus key to the social levelling on display here, whereby economically powerful actors would have to negotiate rights of access and passage with nomadic communities over whom they otherwise ruled. Similar processes were at play on the Senegal River Valley, where the channelling of human mobility also facilitated the extraction of tribute and taxes. In this context, land often constituted the primary resource. However, it was typically owned collectively for subsistence purposes, with taxes in the form of a percentage of harvest being extracted by noble classes. Depending on the time of year and geographic location, this tribute may have taken the form of millet, fish or livestock. In Futa Toro, the floodplain facilitated off-season millet harvesting, making it an attractive place of settlement for many from surrounding regions. The Senegal River was key to the channelling of human mobility and the charging of taxes that went with it; the Fula rulers of Futa Toro regulated right of passage on the river, extracting a tax payment from passing vessels (Yatera, 1996: 22). The lucrative taxation opportunity afforded by control of river traffic was a double-edged sword, however; also subject to such taxes were European traders of slaves, acacia gum, and arms. The nature of this trade eventually placed the European imperial powers in a position of economic dominance, which would soon translate into the political dominance that will be examined shortly.

More immediately, the above should suffice to illustrate the clear qualitative difference between the form of this precolonial channelling of human mobility and the bureaucratic regulation that would come to characterize the colonial era.

There is, however, an equally crucial directional distinction to be made. At this point in time on the middle Senegal River Valley, it was flows moving laterally through the valley on the river that were primarily subject to taxation, rather than those moving vertically across it. The latter patterns of mobility were also common, as many lived on the south bank and cultivated land on the north, but it was not until the colonial encounter that these movements would become subject to regulatory stipulations, as shall be shortly illustrated. As Zekeria Ahmed Salem observes (2004: 19), 'the ways in which residents of the two banks moved in this space has consistently relegated the "separating" function of the river to the back seat'. This observation speaks to the third aspect of territorial organization, namely the mutual constitution of group identity in this precolonial space.

The co-constitution of group identity

We have thus far been proceeding on the basis of what may be a somewhat crude schematic division between *biḍân* nomadic populations of the Sahara and Black African sedentary populations of the Senegal River Valley. While such a breakdown may be useful for reasons of analytical clarity and presentation, the risks of conceding to a colonial geography and obscuring the imbrication and mutual constitution of these overlapping spaces and relations cannot be overstated. This co-constitution applies both within and between the groups of the Sahara and the Senegal River Valley. Concerning intra-group dynamics, the relationship between the two Saharan aristocratic factions, for instance, was a complex and varied one; *zwâya* were often subject to raids and pillaging campaigns or extortionate protection rackets on the part of *ḥassâni* warriors, and in this sense were subservient to them. But a symbiotic relationship also tended to exist between the two, since the ubiquity of Islam as a means of mediating virtually all forms of social relations meant that the *ḥassâni* warriors also depended upon the *zwâya* for scholarly and religious guidance. And while social relations in the Waalo kingdom were stratified between noble *gêêr*, the artisanal *nyeenyo*, and the enslaved *dyaam*, each of these social categories was further divided internally. Such internal lines of differentiation concerned, respectively, issues such as coronary succession, the type of artisanal goods one produced, and whether one had been enslaved through captivity or as punishment (Barry, 1972).

As regards inter-group constitution between the Sahara and the Senegal River Valley, the 1647 Char Bouba war from which the Saharan Emirates emerged in fact had its roots in a religious reform movement that extended deep into both politico-geographic zones (Curtin, 1971). Seeking to gather support for a religious uprising against the perceived corruption of the aristocratic classes, the *zwâya* Berber leader Nāṣir Al-Dîn initially found support amongst the artisanal classes of the Senegal River Valley. The movement enjoyed some temporary success in spurring revolt against the secular *satigi* rulers of Futa Toro, before being defeated and consigned to the region's social memory. In the Sahara, the defeat of the *zwâya* forces in 1677 solidified the subservience of these clerical classes to the *ḥassâni* warrior elite, laying the social foundations of the Emirate system outlined above. But the movement also had a lasting effect on the religious and political constitution of the middle Senegal River Valley. Indeed, for David Robinson (1975), the *zwâya* movement had a determinant role in both emergence of the *torodbe* elite of Futa Toro in the late eighteenth century (Robinson, 1975: 190) and religious and political authority in the Waalo Kingdom (Robinson, 1975: 191).

It is for these reasons that Zekeria Ahmed Salem (2004: 15) suggests the Senegal River Valley be studied from the standpoint of a constitutive unity that is obscured in the present, and that David Robinson (2000) considers this a 'Senegalo-Mauritanian' space (cf. Ould Cheikh, 2004). Abdel Wedoud Ould Cheikh (2004: 114), meanwhile, foregrounds this unity in his analysis of 'the interweaving, simultaneity and reciprocal influence of the itineraries of the chiefs of religious brotherhoods in the region … despite their location on different sides of the border established by colonialism'. This state of constitutive unity and mutual constitution, it must be noted, is a living one. In the villages of Dao and Dolol in the Gorgol region on the middle Senegal River Valley, for instance, the reductive analytical convergence between livelihood, geography, and ethnicity which often informs analyses of Mauritanian society is redundant, given the extent to which economic, cultural, and linguistic lines between *biḍân* and Fulani groups blur into one another (Leservoisier, 2012).

Of course, this is not to say that reductive conflations of racial, geographic, and economic difference are a pure product of the colonial encounter. On the contrary, increased proximity between nomadic Arabo-Berber pastoralists and sedentary African farmers over the course of the seventeenth century resulted in the propagation amongst the former of a discursive dichotomy between 'white' and 'Black' peoples (Webb, 1995).[5] In other words, the idea of a racialized distinction between peoples based on geographical location predates the period of French inland colonization in West Africa (cf. Hall, 2011). The point is not to deny this fact, but rather, as Ould Cheikh (2004: 114) emphasizes, to draw attention to the nature of the territorial formation which sought to enclose these identities. Colonialism was in this sense pivotal not in its fabrication of a racial schism *per se*, nor even in the instrumentalization of an existing one, but in the manner in which this schism was grafted onto an entirely novel territorial arrangement. For while the space of the Senegal River Valley was – and remains – a locus of exchange and amalgamation of difference, the colonial encounter introduced to it a linear conception of territorial division, with all of the racialized overdetermination that this entails.

The colonial shift in territorial organization

Upon disembarking from the Medusa's lifeboat in Saint Louis and taking up his appointment as governor of Senegal in 1817, Julien-Désiré Schmaltz was faced with a no less considerable challenge. Following the loss of Haiti in 1804

5. Cited in Ahmed Salem (2004).

and Napoleon's defeat in 1815, the French Empire was in a position of global weakness during the historical moment captured in *The Raft of the Medusa*. The recuperation of Senegal from Britain formed part of the Treaty of Paris, whose unfavourable terms for the French were felt through Britain's insistence on the suppression of the trans-Atlantic slave trade as a condition of the exchange (Ould Saad, 2004: 92). The social devastation wrought by the slave trade nonetheless created opportunities for the French. In the Waalo Kingdom, for instance, the social contradictions between an elite whose primary tradeable produce was slaves and the land-owning peasantry who constituted the bulk of these export commodities eventually gave rise to civil war and the destruction of the social system. By 1819, French Governor of Senegal Julien Schmaltz was in a position to buy the kingdom and embark upon a project of 'agricultural colonization' (Amin and Girvan, 1973: 187). The Futa Toro region would be facing a similar fate by the mid-nineteenth century. Following recommendations made by a collective of commercial interests based in Saint Louis and Bordeaux, then Governor of Senegal Louis Faidherbe was given the green light for a military offensive on the middle Senegal River Valley in 1854. In response, the ghost of seventeenth century *zwâya* leader Nāṣir Al-Dîn resurfaced in Futa Toro in the form of a *jihâd* against the French, which drew heavily upon the region's legacy of religious reform movements. Nonetheless, by 1891 Futa Toro had been incorporated into the *Afrique Occidentale Française* (AOF) colonial apparatus, laying the foundations for the creation of the colonial territory of Mauritania.

Of course, contact between European colonial interests and the residents of the Sahara far precedes the nineteenth century. The Portuguese took a commercial interest in the Mauritanian space in 1445, building a military base on Arguin Island from which they hoped access could be gained to trans-Saharan trade (Ould Cheikh, 2014b: 15). While relatively fleeting, this endeavour nonetheless stands as example of the broader systemic shift from the Mediterranean to the Atlantic that the era of Portuguese naval imperialism represented (Banaji, 2007: 49). At stake in the following pages, then, is one particular example of what unfolded in the wake of this shift. As is well known, what followed was an era of unprecedented European dominance, operationalized in the spread of capitalism and the linear form of territorial division. In the context of Mauritania, this would entail deep-seated transformations within each of the domains of territorial organization under discussion.

Introducing the border

The representation of the Senegal River as the southern border of the territory comprising the present-day state of Mauritania was first given cartographic expression in 1898, in a plan for colonization submitted to the French Ministry of Colonies. The plan, submitted by colonial functionary Xavier Coppolani, mapped the hostile territories to be brought under French control, which was to be achieved by means of a divide and rule strategy (Coppolani, 1999). Due to the internecine conflict and rivalries that plagued the Trarza Emirate over the course of the second half of the nineteenth century, its reigning *zwâya* were the prime targets of this strategy. An orientalist par excellence, Coppolani drew upon his extensive knowledge of Arabo-Islamic culture and customs in order to convince sections of the clerical *zwâya* classes to cede control of their territories (Désiré-Vuillemin, 1997). Six years later, in October 1904, the 'civil territory' of Mauritania was formally established, in a declaration setting out the first territorial demarcations of the colonial administrative territory of Mauritania. It stipulated that the Senegal River forms the dividing line between the French possessions of Senegal and Mauritania. The city of Saint Louis, where the mouth of the Senegal River meets the Atlantic Ocean, was designated as the departure point of this novel territorial delineation (Ould Saad, 2004). This initial territorial delimitation was given further legal substance thirty years later, through a decree issued on 8 December 1933. It retained and bolstered the Senegal River as the border by providing a greater level of detail about the proprietary status of the myriad islands and backwaters speckled along the river.

As colonial conquest proceeded north and east into the Sahara, the divide and rule strategy that informed the initial colonization did not play out as smoothly. While the relative weakness of certain *zwâya* classes in the south created an incentive to accede to colonial protection, such was not the case for many northern *qabā'il*. As colonization advanced north, it faced military resistance in the form of a *jihâd* spearheaded by Shayk Mâ' al-'Aynâyn, a Sufi scholar and military leader from eastern Mauritania who led resistance to the Spanish conquest of the Western Sahara. As far as territorial legacies are concerned, however, this resistance had ambivalent implications. Mâ' al-'Aynâyn's reliance upon the Sultanate of Morocco for military support laid a foundation for the latter's claim to sovereignty over Mauritania. While evidence of Mâ' al-'Aynâyn's own affinities to the Alawite Sultanate is scant at best (Blalack, 2020a: 13), the legacy of this strategic partnership nonetheless weighed heavily upon postcolonial territorial dynamics, as we will soon see.

The resistance was, in any case, crushed by 1933, and the logic of linear territorial division soon spread throughout the Sahara. The modern-day border between Mauritania and Mali was created by colonial decree on 5 July 1944, significantly expanding the geographical scope of the colony of Mauritania into what had hitherto been French Sudanese territory (Antil, 2004). To the northwest, the line separating the Spanish Sahara from French West Africa had been agreed upon in 1900. But as a line in abstract space with no relationship to the social or topographical features of its representation, it had little bearing on the experience of those it separated at this time (Drury, 2019). It nonetheless provoked low-level geopolitical contestation between these two colonial powers, who reached an agreement in 1957 that would see the peninsula of 'Ras Nouadhibou' – measuring only around 6 km across – vertically bisected by a territorial dividing line. This would afford the Spanish access to the Western side and the rich fish resources off its Atlantic coast. The French, for their part, were ensured access from the coast to the hinterland zones of Idjill and Zouérat, where iron ore extraction operations were getting underway. Again, the intrinsic relationship between mode of production and configuration of space is apparent, in this case with the spread of capitalism dovetailing with the linear form of territorial demarcation (Anievas and Nişancıoğlu, 2015).

While far from an exhaustive overview of the colonial carving up of territory in Mauritania, the above should suffice to illustrate the difference between the precolonial 'itinerant territoriality' discussed in the previous section and the form of territorial distinction introduced through the colonial encountered within space. Cheikh Saad Bouh Kamara and Olivier Leservoisier (2000: 191) give succinct expression to these distinctions: 'while for Moorish tribes, frontiers were open and subject to fluctuations according to the state of political relations, the French sought to impose another conception of space, which entailed the precise physical delimitation of territory'. If precolonial spatial demarcation and distinction were expansive, mobile, and imbricated with overlapping layers of authority and ownership, the colonial variety was based upon neat, schematic assumptions about delineating physical space (Acloque, 2014). It was, in other words, an expression of 'the connection between map making and modern colonialism' (Mezzadra and Neilson, 2013: 33).

Of course, colonial borders in practice rarely operated with the geometric precision by which they were imagined. Borders are, after all, overdetermined by numerous other forces and relations, to use Balibar's (2002) memorable criterion. These forces and relations all come to bear upon how they operate in colonial contexts, as much as in other ones. That said, these colonial borders nonetheless *aspired* to hold a monopoly on territorial distinction in a way that precolonial boundaries did not, and in this respect, marked a significant departure from how territory was imagined to be governed in this space.

For these reasons, an anteriority can be detected within certain aspects of the contemporary border externalization process, insofar as European actors have long been exporting modes of territorial governance in their image and according to their interests. Furthermore, the colonial implantation of the border in what is now Mauritania would act as impetus for radical rearrangements of the other two dimensions of territorial organization, refractions of which can also be seen in the externalization process.

Regulating mobile colonial subjects

The colonial drawing of territorial demarcations immediately posed the problem of how to manage the populations who now crossed these new lines. This is of course the case in any instance of colonial territorial consolidation, but the problem is particularly acute in areas in which the vast majority of the colony's residing population is nomadic. As demonstrated, these nomadic tendencies were a foundational element of organized social life in the precolonial Emirates. They were, however, cast anew by the colonial endeavour. Through the very act of naming and treating them as 'populations', the colonizer bestowed upon itself the mandate of knowing and statistically documenting all those that resided within its territorial purview,[6] an imperative that presupposed a degree of spatial stasis on the part of the population. As Mariella Villasante Cervello and Raymond Taylor (2014: 42) observe: 'the principal change introduced by the colonizer takes place at the level of codification, through registries and surveys.' This was the case not just for the embryonic international borders dividing Mauritania from Senegal to the south and from Mali to the east; the subdivision of the colonial territory into internal administrative units also necessitated the statistical knowledge and documentation of each administrative zone's population.

Three efforts to reconcile nomadic livelihoods with these colonial bureaucratic imperatives can be highlighted. First, at what was now the 'internal' level, *qabā'il* that moved to districts of the Mauritanian territory beyond those in which they had been registered by the French were to be issued with a *laissez-passer* document (Antil, 2004). Information contained in this *laissez-passer* would include the chief of the encampment, the names of the chiefs of its constituent tents, the number of inhabitants of each tent, and the number and type of livestock attached to each tent. Those groups not in possession of a *laissez-passer* would be sent back to their assigned district, whose colonial administrators would be notified accordingly. Another approach, applied in the Gorgol region on the middle Senegal River Valley, was the development of agriculture and property rights through the establishment of a land registry. The logic of this initiative was that the incentive for individual private property would encourage sedentarization and thus facilitate registration of residents of administrative territories (Kamara and Leservoisier, 2000). A ministerial decree was issued to this end on 24 July 1906, offering individuals the opportunity to convert their customary right to a given plot of land into a legal property right, via the land registry system.

Finally, and in direct opposition to the direction in which human mobility was channelled prior to the colonial project, efforts were devoted to regulating the flow of people across the Senegal River. This imperative sprung from the contradictions of converting the river into a territorial line dividing the administrative zone of Senegal from that of Mauritania. Those who had resided on the southern banks of the Senegal River Valley and cultivated land on the north now found themselves subject to two tax regimes – one imposed on residents of the colony of Senegal, and the other on crops cultivated in Mauritania (Leservoisier, 1994). To avoid this double taxation, people began migrating to the north bank in greater numbers. In order to stymy the flow, the colonial government issued an order on 10 January 1905, acknowledging the right of those residents of the south bank who tilled land on the north prior to the arrival of the French to continue to do so without being subject to additional taxes. At the same time, those who had not been engaged in agricultural activity on the north bank prior to colonization but who were now crossing the river to work the land would be subject to both tax regimes.

6. Foucault's (2009) argument regarding the relationship between evolving forms of governmental power and the emergence of 'population' in a metropolitan setting is also evidently applicable to colonial contexts.

These measures aimed at controlling human mobility can be read as a prefiguration of the logic entailed by policies that would follow the externalization of EU migration controls, insofar as human mobility was now seen to represent a problem, whose solution lay in its institutionalization and management. As we have seen, mobility was channelled by elites for extractive purposes in the precolonial era. The difference, however, lies in the bureaucratic quality that the channelling of mobility would take with the colonial encounter. These measures aimed at regulating mobility through statistical knowledge and documentation, which created a path dependency in the realm of migration policy that would prove crucial to the process of border externalization of a later period. Indeed, Gary-Tounkara (2009) highlights how the end of colonial rule in West Africa saw the replacement of the colonial *laissez-passer* document with a national identity card, the latter being a key mechanism by which movement across borders and internally could be judged as legitimate or illegitimate in the eyes of the state.

Racializing territorial belonging

Over the course of the nineteenth century, the mutually constituted nature of the 'Senegalo-Mauritanian' space detailed in the previous section would increasingly be supplemented by a French colonial imaginary that had been developing in tandem with the inland colonization process. By the turn of the century, the Senegal River represented within this imaginary a dividing line between 'Black' Africans who had been brought under colonial rule, and 'Moors' who were yet to be pacified (Ould Saad, 2004: 94).[7] Colonial knowledge production contributed to this racialization of prospective subjects through the depiction of distinct modes of religious practice in operation on either side of the Senegal River: a 'Black' Islam to the south and a 'Moorish' Islam to the north (Robinson, 2000). The 1898 plan for colonization was, furthermore, the first time that the name '*Mauritanie*' was used in reference to this particular geographical expanse of territory, in what Ould Mey (1996: 70) considers to have been a nostalgic allusion to the third-century Roman province of Mauretania, located in modern-day Morocco. By naming this territory with reference to its Arabo-Berber inhabitants – 'the Moors' – the colonial administration formalized the racialized narratives of its own knowledge production regarding 'Black' and 'Moorish' populations. This underlying racial imaginary would structure the 'overdetermination' (Balibar, 2002) of the newly drawn border.

What this meant in practice can be illustrated by returning to the problem of regulating cross-border migration flows from the southern banks to the north banks of the Senegal River Valley. Upon realizing that the colonial order

7. The material reason for the different rates at which the societies of the region came under colonial rule was that the primary trading produce of the Emirates, and that of Trarza in particular, was acacia gum. As Amin points out, the export of gum had significantly less deleterious social effects than exporting slaves.

of January 1905 had failed to stem the flow of Fulani populations to the north bank, the colonial administration adopted another approach. In December 1905, it was ordered that all '*indigènes* of the black African race residing in Mauritanian territory' (colonial administrator's report quoted in Leservoisier, 1994: 62) would henceforth be subject to the same tax regime as that of Senegal. Cleaving a racial distinction within the colonial territory of Mauritania now appeared the most efficient means of preventing the outflow of Fulani populations from Senegal. This policy of racial differentiation is consistent with Mahmood Mamdani's (1996) analysis of late colonialism in Africa, which – he illustrates – relied upon a legal distinction between racialized citizens and 'ethnic' subjects. This distinction often resulted in a dizzying array of ethnically delineated customary legal regimes operating within a single territory, which would, in turn, give rise to a postcolonial state of 'decentralized despotism'. In the case of Mauritania, Bennoune (1978: 37) describes the situation in 1913 as follows: 'the Sub-Saharan Africans living in southern Mauritania were coerced to pay a head tax amounting to four francs per head on every person over the age of ten. The "Moors" resident in the colony had to pay Moslem taxes to the colonial administration comprising the zekkat, one-fortieth of their flocks of herds, and the ushur, one-tenth of their crops.'

From the point of view of the colonial administration, then, cleaving a racial distinction within its territory proved the most efficient means of extracting a surplus from its subject populations. But in effect, the 1905 order rendered the Fulani based north of the Senegal River tax resident in Senegal, despite their physical residence within the territories of Mauritania. This racialized exclusion of Black Africans from the proto-nation-state of Mauritania would resurface in dramatic form over seventy years later, during the 1989 border conflict with Senegal, during which thousands of Afro-Mauritanians were deported from state territory alongside Senegalese nationals. And as we will see in coming chapters, this legacy has been further stirred by the racialized manner in which deportations sanctioned by the EU and executed by the Mauritanian state have been carried out.

As the border form became consolidated across the colonial territory of Mauritania, so too did the understanding of its function as a tool of racial differentiation. This can be seen in an exchange of letters between the governors of the Senegalese and Mauritanian territories and the General Governor of French West Africa (Antil, 2004). The Governor of Senegal believed the territory of Mauritania to be a financial liability, arguing that colonial economic interest would be better served by attaching its more lucrative southern regions to the Senegalese administrative zone, leaving the rest as military outposts. In defence of his territorial pride, the Mauritanian governor emphasized the alleged 'geographical, ethnic and social' distinctions between Mauritania and Senegal (Antil, 2004: 49). He concedes greater degrees of ethnic similarity between the southern regions of Mauritania and Senegal, but nonetheless argues in favour of retaining these areas in order to nourish the rest of the territory financially. Ultimately, the seemingly less pragmatic of the two options wins out, thanks to the overarching imperial goal of ensuring territorial contiguity between the

French possessions of Senegal and Morocco against rival European powers. The eastern border with modern-day Mali, forged by legal degree on the fifth of July 1944, was equally infused with the logic of racial differentiation (Jus, 2003). Speaking of the July 1944 decree establishing the eastern border, Antil (2004: 54) observes that 'the borders of the colony were thus pushed far to the east, so that the greatest possible number of Moors could be incorporated', the aim being 'to place geographical and physical borders between the two races (Moorish nomads and sedentary Sarakollés)'.

It was perhaps only a matter of time before the ethnic categories that informed the process of territorial demarcation filtered into the consciousness of colonial subjects, resulting in the articulation of demands which in a sense mirrored the racialized logic of the colonial border. Cedric Jourde (2004) traces the beginnings of this process to the end of the Second World War, when the tentative rolling out of representative political institutions within colonial territory gathered pace. Over the course of the 1940s, an Afro-Mauritanian, and specifically Fulani, political constituency gradually emerged, initially taking the form of the UGOVF (*l'union des originaires de la vallée du fleuve*). At the time of its establishment in 1946, the UGOVF represented Fulani populations based on both sides of the Senegal River, with the specific aim of countering the perceived dominance of a Wolof elite in Senegalese politics. However, within a period of eleven years, the group would shift the focus of its grievance, shedding its old name in favour of UOMS (*l'union des originaires de la Mauritanie du sud*). As Jourde (2004: 74) notes, 'although the UOMS, just like the UGOVF, is composed of Fulani and concerns itself with the protection of interests in this region, it differs markedly from the UGOVF in the sense that it is the dynamics specific to the Mauritanian arena that define its discourse, strategies and objectives'. Amongst these 'strategies and objectives' were the redrawing of the borders of the territories of Mauritania, so as to 'reunify' the Senegal River Valley. So, while there was a drive to redraw the borders of the Mauritanian state, the strategy to achieve this entailed a greater degree of engagement with this state than had been the case under the UGOVF.

On the eve of independence from colonial rule, the racialization of territorial belonging had been thoroughly imbued within political contestation and social dynamics. An insight into the form this would take in the postcolonial context can be gained from the decision to introduce an Arabic language test in 1952 for prospective colonial civil servants. This decision would foreshadow future contestation around language policies, which Afro-Mauritanian groups would claim function to marginalize them in socio-economic and administrative terms. And while no campaign to implement the UGOVF demand of reunifying the Senegal River Valley ever saw much success, it has been re-articulated by various groups throughout the intervening decades, indicating the appeal that modern ethno-territorial claims now had. These developments demonstrate the extent to which the national model of territorial sovereignty, delivered via the colonial encounter, shapes the nature and content of the ethnically framed territorial counterdemands that emerge within it.

The contradictions of postcolonial national sovereignty

While the political significance of independence from colonial rule should not be understated, it necessarily entailed across the region an institutionalization of rationalities that were introduced by colonialism in the first instance. The years following the implosion of colonial authority in *l'Afrique Occidentale Française* saw the abolition of the *laissez-passer* document, in favour of a new national *carte d'identité*, introduced in 1949 (Gary-Tounkara, 2009). Mauritanian nationality was established by Law 65-112 in June 1961, one year after the country gained independence. The law set out in detail the conditions of what constitutes a Mauritanian national and the criteria for obtaining Mauritanian nationality (Di Bartolomeo et al., 2010). The borders envisioned by Xavier Coppolani at the turn of the twentieth century were now imbued with inclusionary and exclusionary potential and, hence, national validity. In this respect, Mauritania fit within a regional transition in which, as Gregory Mann (2015) has shown, questions about boundaries and the terms of belonging within postcolonial Sahelian states were subject to contestation and debate during the early decades of independence. Henceforth enclosed within Mauritania's national borders were four principal ethno-linguistic groups, namely 'white Moors', or *biḍân*, 'black Moors', or *ḥaraṭîn*, as well as Afro-Mauritanian speakers of Fula, Soninké and Wolof. In what follows we continue tracing the evolution of territorial boundaries, human mobility and group identity. Within each of these dimensions of territorial organization, the agency of the postcolonial state, and strategically situated actors within it, is on display. Equally evident, however, is the manner in which the parameters set during the colonial era continue to condition the postcolonial context.

Disputes and conflict over borders and territory

From the very outset, divergent secessionist territorial claims cast uncertainty upon the borders of the new-born Mauritanian state. At one end were demands coming from Arab nationalist parties *Entente Mauritanienne* and *naḥda waṭaniyya* that the entirety of Mauritania become part of a 'Greater Morocco', as proposed by Allal El-Fassi of the Moroccan *Istiqlāl* party. Such proposals had staunch backing within the Moroccan kingdom, which marked Mauritanian national independence with an official day of mourning. Throughout the 1960s, Morocco challenged the territorial integrity and national validity of the Mauritanian state through international diplomatic fora, like the Arab League, as well as by more covert means, such as through the fomenting of violent dissent in the country (N'Diaye, 2018). Challenging the southern borders of the state were groups such as the UOMS, discussed above, which demanded that southern Mauritania should secede and form an independent territory based around the Senegal River Valley. The task of holding this fragile body politic together was undertaken by Mauritania's first president, Mokhtar Ould Daddah, with the endorsement and active support of the French. Born in the town of Boutilimit in the Trarza region, he was the first child to graduate from the French schooling system that had been progressively installed throughout the colonial territory (Pettigrew, 2007: 79). Of respectable background according to the criteria of *biḍâni* society, while also being schooled in the ways of French culture, language, and bureaucracy, he was well poised to mediate the diverse international and local tensions to which the independent Mauritanian state would be subject.

Ould Daddah adopted several strategies to accommodate and neutralize the territorial pressures that were exerted on the state upon independence. First, the small fishing village of Nouakchott was chosen as the unlikely capital of Mauritania. Despite numbering only 200 people in 1944 (De Chassey, 1978), Nouakchott was seen as a politically neutral location, being situated on the coast roughly halfway between the state's northern and southern borders. Secondly, concerted efforts were made to subordinate the state apparatus beneath the umbrella of Ould Daddah's PPM party, with the state's divergent constituencies and associated political demands put forth as justification for these manoeuvres. These efforts consisted in a constitutional shift to a presidential system of government in 1961, and an (ultimately unsuccessful) attempt to neutralize the fiscal and administrative capacities of the parliament. In preparing the ground for the formal adoption of the one-party model by means of constitutional revision in January 1965, the need for national unity and rising above 'particularisms' was consistently invoked (Marchesin, 2010: 120). Indeed, the now-ubiquitous trope of Mauritania as a 'hyphen' or 'bridge' between two otherwise distinct geographical, cultural, and ethnic spaces was produced and actively disseminated by the ruling party at the time of this transition from a multiparty system to one-party rule (De Chassey, 1978: 290).

The drive to cement the link between national belonging and territorial control would eventually take an expansionist turn, however, ultimately bringing about Ould Daddah's downfall. Spain's withdrawal from Western Sahara in 1975 was seized as an opportunity to push Mauritania's borders outward to encompass this territory whose inhabitants were, in the discourse of Ould Daddah, culturally, geographically, and religiously, part of Mauritania. Crucially, alongside the familiar motifs of national unity and territorial integrity, the 'artificiality' of the north western border was invoked as a justification for this territorial expansion (Marchesin, 2010: 148). What comes under challenge in this postcolonial context, then, is not the linear border form itself, but rather the question of where it ought to be located, and who ought to be positioned on what side of it. There is thus a peculiar resilience to this form of territorial distinction, which is explicable not just by its effectiveness in subjugating groups to the bureaucratic and surplus extraction imperatives of the colonizer, but also in its capacity to be appropriated by some of these same groups to enhance their own share of power in the postcolonial context. The north Mauritanian border embodied this paradox since, as Mark Drury (2019) argues, it was only at the moment of decolonization that its capacity to dictate allegiance and regulate movement truly kicked in.

The potency of the northern border is a direct result of the multiple conflicting territorial claims that were being made upon it at the moment of independence. These claims were initially voiced by Mauritania and Morocco at the International

Court of Justice, which was petitioned to offer clarity on the slippery question of precolonial links between the Western Saharan territory and the spaces of Morocco and Mauritania (cf. Blalack, 2020a). But the ambiguity of the ruling[8] meant that each actor could play it to their advantage. The infamous 'green march' led by Moroccan King Hassan II into the Western Saharan territories set the stage for a more confrontational approach than the diplomatic one favoured by Ould Daddah up until that point. A subsequent increase in geopolitical hostility and militarized resistance on the part of the Saharawi populations dragged the deeply unprepared Mauritanian armed forces into conflict in 1976. The misfit between this expansionist endeavour and the third wordlist anti-imperialist rhetoric of the Mauritanian government was not lost on many both at home and abroad. Moreover, the economic pain inflicted by Polisario attacks on iron ore transport networks generated increasing disillusion amongst the Mauritanian population. In July 1978, a small group of colonels capitalized upon this growing disenchantment by ousting Ould Daddah in a bloodless coup.

During the ensuing era of military rule in Mauritania (which some would argue to be ongoing[9]), the borders of the state enjoyed a greater level of stability than that which had characterized the opening years of independence. While the Western Sahara adventure was a disaster for many in Mauritania, not least Ould Daddah himself, it may also have finally established the legitimacy of the Mauritanian state in the eyes of Morocco, thereby relieving some of the territorial pressure coming from the north of the country. Moreover, with the ousting of southern *zwâya* leader Ould Daddah, the way was paved for a succession of leaders from warrior *qabā'il* to take the reins of the state (N'Diaye, 2018: 34). This to a certain extent dampened the social contradictions wrought by the colonial encounter, by shifting state control in favour of those northern groups who had been violently subjugated during the colonial era. This reorientation of state power from the south-west to the north thus deepened the territorial consolidation of the postcolonial Mauritanian state.

While there continued to be a level of ethnically based secessionist contestation in the south of the country, these claims were for a period overshadowed by a framework of regional cooperation on the Senegal River Valley. The first regional agreement regarding the development of the Senegal River Valley dates to 1963, but it would only be with the creation of *l'Organisation pour la mise en valeur du fleuve Sénégal* (OMVS) in 1973 that any progress in infrastructural development would be made (Ould Saad, 2004: 100). Bringing together the states of Mauritania, Mali, and Senegal, the OMVS was intended to address coordinate and integrate collective responses to regional challenges, such as drought and demographic pressures (GRDR, 2014: 20). This it achieved through financing and supporting agricultural

8. The ICJ established juridical links between each of these territories in the precolonial era, but not that territorial sovereignty had been exercised by either Morocco or Mauritania.

9. Mauritania transitioned to a multi-party democracy in 1991, but has been consistently presided over by high-ranking military officials who have retired from the ranks to run for office (N'Diaye, 2018).

modernization schemes as well as through two dam construction projects, which lasted between 1982 and 1988. In facilitating cross-border cooperation between these three postcolonial states, the OMVS would offer a temporary resolution to the social contradictions engendered by the colonially implanted border. It was, moreover, but one of several overlapping regional cooperation frameworks, the most notable of which is the Economic Community of West African States (ECOWAS, to which we return below), each of which simultaneously reinforced the borders of West Africa while simultaneously reconciling some of their more antagonistic origins. Key to this reconciliation was the relative ease with which nationals of ECOWAS member states could travel between countries, which brings us to the second dimension of territorial organization under consideration in the postcolonial context.

Labour migration and social transformation

The period of independence in Mauritania was characterized by a radical shift in the nature of human mobility and the socio-economic context in which it occurred. This shift consists in a gradual but secular decline of the nomadic population and a concomitant increase in migration to cities, the latter taking both internal and transnational form. At the time of independence in 1960, the country's residing population numbered around 2 million people, around 75 per cent of whom were nomadic. By 1976, however, the nomadic population had dropped to 27 per cent, while the urban population had increased to 31 per cent, from 6 per cent in 1959 (Bennounne, 1978: 46). By 2015, the urban population had grown to over 1 million, representing 48 per cent of the national population, of whom just 1.9 per cent were still nomadic (Islamic Republic of Mauritania, 2015: iv). West African migrants would make up increasingly large portions of these new urban populations. By 1988, there were 51,330 foreign nationals residing in Mauritania, 81.6 per cent of whom originated from Senegal, Mali, and Guinea Conakry, representing 2.75 per cent of the residing population (Di Bartolomeo et al., 2010). Emigration would be primarily oriented towards West Africa, although a smaller proportion of Mauritanians from the south of the country, primarily Soninké, would migrate to France in the 1970s and 1980s (Yatera, 1996).

These migrations took place in a contradictory legal environment. For, on the one hand, the exclusionary boundaries of the Mauritanian nation-state were given formal expression in 1965 through the passing of Law 65-046, which stipulated that foreign nationals who entered the country irregularly would be punished with a two-month to two-year prison sentence (Di Bartolomeo, Fakhoury and Perrin, 2010). On the other hand, however, this remained a rather paltry legal framework for much of the period of independence. It was not until 2008 that legislation would set out in any detail the conditions of access to employment for foreign nationals, and there is as yet no legal stance on family reunification. Furthermore, Mauritania's membership of the Economic Community of West African States (ECOWAS) meant that, from 1973, nationals of its member states were able to circulate freely within the bloc's territory without being subject to

Table 1 Breakdown according to the nationality of foreign nationals residing in Mauritania (Islamic Republic of Mauritania, 2010a)[10]

Country of origin	Number of nationals in Mauritania
Senegal	35,000
Mali	16,000
Guinea	30,000
Guinea Bissau	300
Togo	600
Côte d'Ivoire	750
Central African Republic	150
Gambia	1,600
Total	**84,400**

visa requirements. While Mauritania withdrew from ECOWAS in 2000, it retained many of these bilateral agreements pertaining to free movement. For these reasons, for much of the history of the Mauritanian state, the regulation of migration fell beyond the purview of legislative implementation and, for the most part, regulated itself informally (Bensaâd, 2008). All of this mirrors the regional state of affairs, embodied in ECOWAS, which did not so much provide new legal pathways to migrate as institutionalize pre-existing informal mobility patterns (Idrissa, 2019; Arhin-Sam et al., 2022).

While migration has been paid little legislative and public attention throughout the history of the Mauritanian state, it has also played a structural role within this history. Consider, by way of example, a particular northern *biḍân qabîla*, named the *awlād bousbā'*. At the moment of independence, those members whose families had been engaged in the trans-Saharan caravan trade formed part of a regionally dispersed business network which proved highly adaptive to a nascent mode of capitalist production in the country. Writing elsewhere, Bensaâd (2011) lays particular emphasis on how this capacity for regional mobility, inherited from the trans-Saharan caravan trade, has worked to the advantage of such 'globalized tribes', which tend to hold positions of particular influence and proximity vis-à-vis

10. This data is of course somewhat outdated. More recent statistics are offered in a Mauritania migration profile published by the United Nations University and Maastricht Graduate School of Governance (UNU-Merit, 2017), which are drawn from the UNDESA international migration stock 2015 data. This places the overall number of foreign nationals in Mauritania at 138,162. It suggests Malian nationals to have overtaken Senegalese nationals as the most populous foreign national community in the country (56,557 and 38,574, respectively), a result of the conflict in Mali that broke out in 2012. However, there appear to be significant gaps in the UNDESA data set as it does not include any data for nationals of Gambia or the Central African Republic. The figure provided for Guinea (3,105) is also likely to be less accurate than the 2010 figure offered above in Table 1.

the Mauritanian state.[11] Another pertinent feature of the *awlād bousbā'* business model is the role that migrant labour has to play within it. Tied up in their transport industry is the buying and selling of engines and other car parts. In order to handle this aspect of their affairs, members of the *awlād bousbā'* often turn to Senegalese migrant labour. Within the Le Ksar district of Nouakchott today one can find a bustling district full of shops that deal in the sale and repair of car parts, overwhelmingly staffed by migrants from Senegal. For this reason, Bensaâd (2011: 27) argues immigration into the Mauritanian state to be structurally linked to the emigration of Mauritanians to other parts of the region. This dynamic between *biḍân* owners of capital and West African migrant labourers is, moreover, replicated across numerous sectors of the Mauritanian economy.

This dynamic extends beyond the scale of the newly established Mauritanian state and the region of West Africa, however. Indeed, it is bound up in the global centre-periphery asymmetries that formed the backdrop of the colonial shift in territorial organization detailed above. How these dynamics played out in the immediate postcolonial context is exemplified in the port city of Nouadhibou in northern Mauritania. While recently associated with illegalized Europe-bound migration (e.g. Tuqoui, 2006), Nouadhibou had in earlier decades served as a node through which raw materials extracted in Mauritania could be exported to Europe. The discovery of iron ore minerals in the north of Mauritania at the twilight of the colonial era in 1952 resulted in the transformation of the remote northern settlement of Zouérat into a mining town. It would soon be home to a handful of well-paid European mining officials and a much larger number of low-paid workers of Mauritanian and West African origin (Bennoune, 1978: 42). Mining operations were overseen by MIFERMA (*Société Anonyme des Mines de Fer de Mauritanie*), a conglomerate of European steel interests. Export of raw materials to Europe was facilitated by a railway that was laid connecting Zouérat to Nouadhibou, where construction of a large port was underway. This harnessing of domestic infrastructural initiatives to the interests of European capital is emblematic of the centre-periphery dynamic, insofar as raw materials could be extracted and exported from the Mauritanian periphery at a production cost lower than that available in the European centre (Amin, 1974a: 13).

This modernization drive necessitated a degree of labour power that was largely not to be found within the domestic population, which, as alluded to above, overwhelmingly consisted of nomadic pastoralists and subsistence farmers in 1960. The decades following independence therefore saw Nouadhibou transformed into a regional pole of labour migration (Choplin and Lombard, 2008, 2013; Poutignat and Streiff-Fénart, 2008), with people migrating from various parts of West Africa to work in a nascent industrial fishing industry, the transport sector, and as plumbers, construction workers, and electricians. These migrations

11. Former President Mohamed Ould Abdel Aziz, who stood down in August 2019, exemplifies this tendency. He is a member of the *awlād bousbā'* and is widely known to have been born in Senegal, due to the commercial activities of his family in the country.

are thus part and parcel of the infusion of urban capitalist social relations within communities and societies characterized largely by nomadic pastoralism and sedentary agriculture, a transformation engendered by the MIFERMA mining operation's cementing of the postcolonial Mauritanian state's peripheral economic relationship to the European capitalist centre (Amin, 1973: 73). Of course, the days in which European capital was the sole economic actor have long past, and with it Mauritania's relationship of exclusive dependency vis-à-vis Europe. As we will see in the next chapter, however, the geographic imaginary of the EU border regime's external operations has revived this centre-periphery frame (cf. Casas-Cortes and Cobarrubias, 2019).

These recent EU border externalization policies have also politicized migration in a fashion that represents a marked departure from the immediate postcolonial context under consideration here. This is not to say, however, that this context was free of other forms of exploitation and racialized marginalization to which West African migrants were frequently subject. In other words, while the framework for regulating human mobility that the colonizer introduced remained rather dormant in the decades following independence, the racialization of territorial belonging would prove rather more salient in this era.

Racialized territorial belonging in the postcolonial era

The exclusionary potential of each of the three dimensions of territorial organization would converge at the end of the 1980s, during a period of racialized violence involving mass deportations from Mauritanian state territory. What has since come to be described in Mauritania and Senegal as 'the events' of 1989 has many roots in the colonial racialization of territorial belonging outlined above, but it also built upon processes that developed in the late colonial era and opening decades of independence. The legacy of French educational policies in the colony of Mauritania meant that, for a time, Afro-Mauritanian populations were most suited to administrative positions, due to their greater degree of integration within Francophone culture (cf. Pettigrew, 2007). However, the early decades of independence would see a gradual shift towards the country's *biḍân* constituency, with a trend of 'Arabization' becoming prevalent.

An early example of this tendency and the social tensions it generated can be found in ministerial decree 66.004, passed on 13 January 1966, which made Arabic a compulsory subject in school. Afro-Mauritanian groups organized rallies and strikes in protest, to which Ould Daddah responded by imposing a curfew and deploying the military (N'Diaye, 2018: 47). Arabization would gain further traction after the first of Mauritanian's many coups in 1978, and the ascendance to power of the armed forces. This is illustrated by Antil and Lesourd (2012), who show that between 1958 and 2011, 75 per cent of ministerial portfolios went to members of the *biḍân*, with the remaining 25 per cent distributed between members of the Fulani, Soninké, and *ḥaraṭîn* communities. In the early years of independence at least, this is evidence of certain beneficiaries of the colonial era extending their privileges into the postcolonial context. At the same time, however, it is a reaction against a perception of colonial dynamics being perpetuated through the education and administrative system, given it entailed a marginalization of those perceived to be more proximate to Francophone culture.

That said, the Arabization trend did not go unchallenged, with alternative perspectives on 'the national question' taking multiple forms. Following a bloody crackdown on striking mine workers at MIFERMA in 1968, soul searching amongst radical intellectuals, students and civil servants led to the formation of a multi-ethnic grouping calling itself the *Mouvement National Démocratique* (MND). Out of this revolutionary movement a party was formed in 1973 calling itself the *Parti des Kahidines de Mauritanie* (PKM).[12] The socio-economic analysis of the MND and PKM laid emphasis upon the 'neo-colonial' character of the Mauritanian state, the subjugation of the local bourgeoisie to the dictates of imperialism, and the unified position of 'the people' in opposition to global imperialism and its local enablers. As Francis De Chassey (1978: 406) states, this analysis casts a different light on the national question. Indeed, the MND in its early years decried 'Arab chauvinism' and called for equal recognition of the country's four ethno-linguistic groups. It also immediately lent its unequivocal support to the Polisario Front's struggle for self-determination in Western Sahara (De Chassey, 1978: 406). Grievances and demands specific to the Afro-Mauritanian community were also articulated by an assortment of groups and movements of both moderate and far-left persuasion. Many of these came together in 1983 beneath the banner of *Forces de liberation des Africains de Mauritanie* (FLAM). Composed primarily of intellectuals and members of the civil service, this movement challenged the Arabization agenda in a fashion that, according to some, mirrored Arab nationalist chauvinism (Marchesin, 2010: 329). With the onset of the neoliberal era and a deepening ecological crisis, the anti-imperialist cross-community perspective would lose ground to these largely incompatible ethno-national visions.

The monopolization of the state by a *biḍân* elite coincided with a series of reforms that liberalized the Mauritanian economy in the 1980s and 1990s, orienting it increasingly towards foreign investment (Ould Mey, 1996). The Mauritanian variant of neoliberalism also saw an overhaul of the traditional system of land ownership, through a new land law introduced in 1983, which abolished the system of system of collective ownership that had existed up until that point (Leservoisier, 1994). In its stead, a system based on individual private property rights was introduced. Combined with intense droughts in the 1970s and 1980s which pushed many nomadic pastoralists to the Senegal River Valley (cf. Meillassoux, 1974; Sen, 1983), this law facilitated the acquisition of large tracts of land by some of the more well connected of these pastoralists (OECD, 2010). All of this placed strain on the political relations of the OMVS which had temporarily dampened the social contradictions of the region's colonially inherited borders, as noted above. These tensions would be further sharpened following a coup attempt

12. 'The Mauritanian Proletarian Party'.

in 1987 by Mauritanian officers affiliated to FLAM, and the execution of those identified as the ringleaders by the military (Santoir, 1990).

Together, these processes set the stage for a spate of ethnically framed violence that quickly escalated into a military standoff between Senegal and Mauritania. What started out as a local dispute between Mauritanian Fulani herders and Senegalese Soninké farmers degenerated with dramatic rapidity into communal violence in both Mauritania and Senegal. In Mauritania, this took the form both of vigilante reprisals and organized state deportations. And in each case, Afro-Mauritanians were often lumped in with Senegalese nationals. Over the course of what came to be known as 'the events', 95,000 Black Africans were deported to Senegal, including 25,000 Afro-Mauritanians. 170,000 Mauritanian *biḍân* also fled Senegal to Mauritania (Parker, 1991). As mutual expulsions gathered pace, the border between the two countries was closed and diplomatic ties severed. The kernel of racialized territorial belonging implanted during the colonial era had come to the surface in Mauritania, and with it the contradictions between the national model of territorial sovereignty and the colonial context from which it emerged. In the next chapter, we will see how the externalization process has become entangled within this history, through the mass deportations and racialized policing of migration that it impels.

Conclusion

'Mauritania', Abdel Wedoud Ould Cheikh (2014a: 9) observes in the introduction to a volume commemorating fifty years of Mauritanian independence, 'constitutes, in the main, a French territorial creation'. Of course, he is quick to point out, this does not mean that the French created the people who inhabit this territory, nor that their religious practices or myriad group affiliations were substantially modified over the course of colonial rule. On the contrary, these emerged rather unscathed, relative to the colonial experience of other African states. Paradoxical though it may sound, the very artificiality of the territorial borders drawn during the colonial era may have facilitated the resilience of the multiple layers of group affiliation which transcend them. In this light, does an analysis that emphasizes the colonial experience not bolster this artificiality, endowing it with an undue validity? Such risks dwell beneath the surface of this chapter, which did not, after all, touch upon numerous predecessors to the contemporary Mauritanian state, such as the Takrur Empire, the Ghana Empire, the Almoravid, and Almohad Empires (cf. Al-Naqar, 1969; Norris, 1969; Ould Cheikh, 2018; Pettigrew, 2019). There are thus multiple sedimented social layers embedded in the contemporary Mauritanian nation-state which make up a silent backdrop to this chapter's focus.

That said, in terms of the specific analytical aim to which this chapter contributes, it would be ill-advised to treat the colonial encounter as any other period of Mauritanian history. As we have seen, the colonial era in Mauritania involved a qualitative shift in territorial organization, which we detailed from the perspective of the evolution of three dimensions of territorial organization. In this light, the colonial externalization of the border form of territorial distinction to Mauritania comes into view. So too does the bureaucratic regulation of human mobility that emerged in the colonial era, which represented a marked departure from Saharan nomadic pastoralist patterns of movement as well as those of the precolonial Senegal River Valley. Meanwhile, the racialization of territorial belonging created a reductive administrative dovetail between pre-existing ethno-racial groups and an entirely novel linearly demarcated territorial formation. These developments would sharply condition the nature and form of the independent Mauritanian nation-state's trajectory, a process to which postcolonial state actors were key. This can be seen in Ould Daddah's justification for the Western Saharan adventure, which invoked the 'artificiality' of the northern border while simultaneously upholding the French colonial assumption of a congruence between ethno-national identity and the linear border form. It is furthermore evidenced in the Arabization trend which sought to further cement this congruence. The postcolonial era is thus characterized by a dual process – the colonial conditioning of the postcolonial context, on the one hand, as well as the assumption by postcolonial state actors of steering roles in the maintenance of this state of colonial conditioning, on the other. This is above all evident in the racialized expulsions of 'the events' of 1989.

With the introduction of EU bordering prerogatives in Mauritania from 2006, this legacy would be stirred once more. Indeed, the externalization process has resulted in the highest number of deportations since 'the events', with 11,637 taking place that year alone (Di Bartolomeo et al., 2010). Despite obvious contextual differences, the fact that expulsions from state territory have once again become common practice immediately brings forth the three domains of territorial organization, and in particular their exclusionary convergence as it is embodied in the act of deportation. More generally, the externalization process that has impelled this more recent bout of deportations is overlain by the geographic imaginary of the Mediterranean centre-periphery boundary, which, as we saw in the previous chapter, has sat atop the development of the EU migration and border regime. As externalization advances, however, this geographic imaginary appears to fragment and contract, suggesting its manifestation in the present to be an expression of a decline in the dominance that facilitated the colonial shift in territorial organization.

Chapter 4

EU Border Externalization in Mauritania

At the Chamber of Commerce in Nouakchott, delegates gathered in July 2018 for a two-day conference on migration and development. Included amongst the attendees were the IOM's head of mission, the head of the European Delegation, the president of the University of Nouakchott, several members of its arts and humanities faculty, numerous heads of migrant community associations based in Nouakchott, and civil servants from the Ministries of Justice and Interior. A packed programme kept delegates captive in the hall for most of the two days, but periodic opportunities to rub shoulders were offered in an adjacent room over tea, coffee, and one lavish lunch. Clustered along the walls of this lunchroom were stalls hosted by a variety of international organizations and Mauritanian civil society associations, their tables laden with brochures and pamphlets pointedly situating their line of activities in relation to the conference theme. Here, new networks and relations were forged and pre-existing ones renewed, in some cases through a renewal of antagonism. A prominent figure from the Sierra Leonean community took an IOM staff member aside to have a word about a case in their community. A Ministry of Justice official sipped coffee and perused the pages of a report published by a Mauritanian NGO condemning the Mauritanian government. Graduates from the new master's programme in Migration and Development at the University of Nouakchott hopefully pitched their theses to anyone who would listen.

Organized by academics at the University of Nouakchott and financially supported by the EU support programme for the national migration strategy, the conference was a microcosm of the constellation of actors, relations, and practices that have come to constitute the EU border externalization process in Mauritania. This is so in terms of the social networks reproduced at such events, the policy vocabularies and priorities naturalized and disseminated within them, the asymmetric relations that structure them, and the mediating agents that sustain and reproduce them. This chapter will outline the events and processes that resulted in this constellation of actors adopting such a devoted concern with migration. It will identify how such extravagant resources came to be committed to an issue which had, in recent memory, scarcely entered the realm of policy consideration in Mauritania. Most importantly, it situates these events, processes, and actors within the history narrated in the previous chapter, and in particular the evolving geographic imaginary of the Mediterranean centre-periphery boundary.

From this vantage point, the chapter argues the externalization process to be marked by a dynamic tension; on the one hand, externalization's colonial conditioning is evident, both in its general reinforcement of power relations whose lineage lies in the colonial era, and in its more specific bolstering of the colonial shift in territorial organization detailed in the previous chapter. On the other hand, however, the Eurocentric geographic imaginary of the Mediterranean centre-periphery boundary, which is so foundational to the Eurocentrism upon which the EU border regime is premised, appears to be contracting as externalization proceeds. This can be seen in Mauritanian's repositioning within this geographic imaginary from the West Mediterranean to the Atlantic Route, and more recently to the Sahel. This is not to say that the colonial dynamics examined in the previous chapter that emerged from the erection of the Mediterranean centre-periphery boundary are contracting in tandem. On the contrary, they are in certain respects intensifying, as we will soon see. But there can nonetheless be discerned from the vantage point of Mauritania a reconfiguration of the Eurocentric imaginary of the Mediterranean centre-periphery boundary that is being driven by the externalization process.

The chapter develops this point in three, more or less chronological, stages. The first details the initiation of the externalization process in Mauritania in 2006 following the arrivals from West Africa on the Canary Islands over the course of that summer. What followed was a suite of externalization measures, initially aimed at stemming the unwanted arrivals and then at embedding this newfound policy concern with migration into the architecture of the Mauritanian state.[1] In the process, externalization's strategic logic of containment expands from 'hard' security measures to encompass 'softer' developmental ones, situating it within a more general 'liberal way of development' (Duffield, 2010). As it proceeds into this latter phase, however, externalization becomes marked by inertia. The second section details those elements of externalization that overcame this inertia to see a significant degree of project completion, explaining this uneven implementation with reference to the three dimensions of territorial organization that oriented the previous chapter. In this fashion, the externalization process is shown to be conditioned by the colonial shift in territorial organization outlined in the previous chapter.

While the material and ideological asymmetries of the Mediterranean centre-periphery boundary are thus at play in the externalization process, these asymmetries have been significantly rearranged in the wake of the arrivals to Europe in 2015 and the consequent launch of the EU Trust Fund for Africa. The final section details two elements of this rearrangement; the first concerns a further shift in the EU's geographic imaginary indicated in the Trust Fund, which has expanded the Sahelian periphery at the same time as the Mediterranean centre-periphery boundary has

1. In narrating this transition from defensive to preventative dimensions of externalization, I take heed from scholarship that has analysed EU external migration and border governance in these terms (Boswell, 2003; Triandafyllidou, 2014; Vives, 2017).

further contracted. The second concerns externalization's paradoxical relationship to the far-right and authoritarian politics, which it bolsters while simultaneously justifying itself on grounds of stemming populism in the European electoral arena. Together, each of these is indicative of the era of dwindling European supremacy that is expressed in the externalization process.

'Hard' and 'soft' externalization in Mauritania

The foundations for events such as the conference discussed above were laid in Mauritania in the closing months of 2005 and early 2006. In October 2005, a brutal and fatal standoff between migrants and security forces at the Ceuta and Melilla border fences led many of those on the move to Europe to look for more viable departure points farther south, turning in many cases to the Mauritanian port city of Nouadhibou (Goldschmidt, 2006; Choplin and Lombard, 2013). As externalization developed as the prime method of response, Mauritania was framed within Frontex knowledge production systems as a crucial partner state within the operational area of the Atlantic and the Canary Islands (Frontex, 2010: 84). This positioning of Mauritania on the Atlantic coast of West Africa marks a subtle departure from prior locations within the EU imaginary of external migration governance. Indeed, as far as migration control is concerned, Mauritania has historically been framed as a Mediterranean partner state. This goes as far back as the 5+5 dialogue, which was launched at the 1990 Western Mediterranean Forum with the aim of tightening security cooperation between five Mediterranean states north of the EU external border and the five Maghreb states (European Commission, 2022). While the forum's spatial imaginary of the 'Western Mediterranean' has re-emerged in Frontex's designated zones of intervention, Mauritania has shifted beyond this operational area, being defined instead according to its West African Atlantic location. Mauritania, in other words, moved outside the Mediterranean centre-periphery boundary with the launch of the externalization process off West Africa in 2006. As we will see, this geographic imaginary has contracted further since.

Negotiating the defensive edge of externalization in Mauritania

Over the summer of 2006, the tourists who ritually flock to the beaches of the Canary Islands every year were accompanied by less fortunate visitors, many of whom had arrived after travelling upward of 700 kilometres across the Atlantic in small fishing boats. Over the course of what came to be called the '*cayucos* crisis' in Spanish media,[2] the Spanish Government recorded 31,863 'irregular' arrivals on the Canary Islands (Carrera, 2007: 13). While this would represent no more

2. Named after the Spanish word for the vessels in which people were arriving. Known as *pirogue* in French, media often also make reference to the '*pirogue* phenomenon' (Mayault, 2017).

than 5 per cent of arrivals to Spain that year (Migreurop, 2010: 2),[3] the border spectacle (De Genova, 2013a) it generated was nourished by politicians and media alike. After a Spanish government request to the EU to intervene, Frontex's maiden naval mission was launched. Dubbed Operation Hera I, it involved a simple seventeen-person deployment on the Canary Islands. But the focus soon drifted offshore after another Spanish government request in June resulted in the launch of Hera II, whose aim was to dissuade boats from leaving the West African coastline (Wolff, 2008: 258). This was a significantly more elaborate operation, consisting of naval and air patrols deployed to intercept boats, at a cost of €3.5 million (Migreurop, 2010: 13). The possibility of lodging an asylum claim in Spain would only be available if interception occurred beyond a zone stretching 24 miles from the coast.[4] The operational logic and coordination of what has since come to be termed EU border externalization was now set in motion, with the shores of West Africa in 2006 acting as the testing ground.

Like so many other contexts, externalization here involves a peculiar asymmetric operation of the law on either side of the EU external border. For, on the one hand, deportations from Spanish state territory to Mauritania are legally grounded in a bilateral readmission agreement signed between the two countries in 2003. This created a legal framework in which the deportations from Spain to Mauritania that skyrocketed in 2006 could occur. That said, this framework is itself infused with geopolitical asymmetry; under its terms, Mauritania is obliged to readmit not only its own nationals, but also individuals deemed to have transited through Mauritanian state territory to reach Spain (Strik, 2010: 11). Legal precedent for these skewed terms is, moreover, provided in the ACP-EU framework agreed at Cotonou in 2000,[5] whose Article 13 readmission clause not only commits contracting parties to the readmission of their own nationals, but also makes provisions 'for the readmission of third-country nationals and stateless persons' (Strik, 2010: 2). While this clause is clearly a reflection of geopolitical power relations, it is a legal clause nonetheless.

3. Indeed, while irregular arrivals represented only 5 per cent of the total number of arrivals, the number of irregular arrivals by fishing canoe on the Canary Islands in 2006 was a much smaller proportion still. The 603 *pirogue* arrivals (Carrera, 2007: 13) was an increase on the 214 recorded the previous year, but it is a minute fraction of the 31,863 irregular arrivals in 2006.

4. Hera was renewed for a third time in 2006, at the marginally lower cost of €2.7 million. The years 2008, 2009, and 2010 have each also had a dedicated Hera Operation, the 2008 Operation lasting 315 days and the latter two lasting 365 each (cf. Migreurop, 2010: 13).

5. The Africa Caribbean Pacific Partnership–European Community agreement, signed in Cotonou in June 2000, is an agreement between the EU and the seventy-seven states of the region it designates. It offers a framework in which relations are conducted between the EU and these states. The European Development Fund (EDF) is the main body through which aid and other financial instruments are allocated.

On the other side of the borderline, however, the law appears to fade into obscurity. This was made clear following the establishment of Mauritania's first migrant detention centre in Nouadhibou in April 2006 (Global Detention Project, 2021). In an investigative report on the situation of migrants deported from Mauritania, Amnesty International (2008: 22) notes 'the absence of any legal character for this detention centre', whose creation seems to have emerged from a high-level ministerial meeting held on 16 March 2006 between the Spanish Secretaries of State for Foreign Affairs and Security and the Mauritanian Ministers of the Interior and for Foreign Affairs and Cooperation. This legal opacity soon earned the centre the nickname *Guantanamito*, or 'little Guantanamo' (Choplin and Lombard, 2013: 74), a name it lived up to in its practices as well as its origins. Indeed, of the 4,500 people who detained and deported via *Guantanamito* in 2006, most had every legal right to be in Mauritania (Panzeri, 2010: 12), since existing migration legislation did not criminalize leaving the country (Maccanico, 2010: 2).

This spike in detention and deportation involved a deep restructuring of the social fabric in Nouadhibou. While the outsourcing of responsibility for detaining and deporting unwanted migrants largely succeeded in attaining its goal of reducing arrivals in Spain, this came at the cost of a level of state deportations in Mauritania not seen since 'the events' of 1989–91. The increase in detainees also had a similarly deleterious effect. While the number of migrants detained in the Canary Islands dropped from 31,678 in 2006 to 2,246 in 2009, 'the number of people detained [in Mauritania] on the basis of "suspicion" of wanting to reach Spain has remained stable at between 300 and 360 people per month' (Maccanico, 2010: 2). The cost of the 'success' of these defensive externalization operations has thus been Nouadhibou's conversion into a bottleneck in which migrants could accumulate, to be dispersed across the region at periodic intervals (cf. Choplin and Lombard, 2013).

If centre-periphery asymmetries between the EU and Mauritania are observable here, new political leverage opportunities have also emerged within the Mauritanian periphery. Indeed, the mounting drive in the Global North to prevent unwanted migration has generated a new form of 'migration diplomacy' (Adamson and Tsourapas, 2019), whereby certain Global South states find themselves in a position of relative diplomatic influence in light of their ability to facilitate or restrict the movements of 'undesirable' people (cf. Ferrer-Gallardo and Gabrielli, 2022). This was particularly so in Mauritania in the mid-2000s due to the nature of the domestic political juncture in which the externalization measures outlined above unfolded. Mauritania was at this time undergoing a period of fraught transition, after the ousting by coup d'état in August 2005 of President Ma'aouiya Sidi Ould Taya, who had ruled the country since 1984. The coup generated a crisis of legitimacy in the eyes of the international community, whose concerns about its violations of liberal democratic creed were most vociferously voiced by the EU.

To address these concerns, the European Commission made a call for formal consultations to open between the EU and the newly established Military Council for Justice and Democracy (European Commission, 2005). Aid would only continue to flow if the conditions of its associated financing agreements were viewed to have been met. This was the state of relations between the EU and Mauritania when arrivals began to increase in the Canary Islands in late 2005 and early 2006. A democratic presidential election in March 2007 satisfied Cotonou pre-conditions for the resumption of aid flows and relations, but it is easy to envision how in the interim, the transitional military government might have found itself with additional leverage as a result of the national panic that was unfolding in Spain. Indeed, the EU-Mauritania Fisheries Partnership Agreement (FPA) signed at the height of the '*pirogue* crisis' in June 2006 was the most expensive fisheries agreement the EU had ever signed with a third country at that point (Migreurop, 2010: 13).

The role of such bargaining chips within normalization negotiations was made plain when another coup struck in August 2008. This time led by Mohamed Ould Abdel Aziz, the 2008 coup upended the democratic transition process that had been initiated in 2005, setting the ACP-EU clock of liberal state building back to zero. To revive the normalization process, a European Parliamentary delegation for relations with Maghreb countries led an assessment mission to Mauritania between 15 and 18 February 2010. The investigative report produced by this delegation (Panzeri, 2010) addresses four areas the EU viewed to be of priority for the resumption of normal diplomatic relations: two pertaining to the domestic sphere,[6] and two international issues of 'mutual interest' – namely migration and fisheries. The comments made by Mauritanian Prime Minister Ould Laghdaf to his European counterparts about each of these two read like a case study in the practice of migration diplomacy. As regards migration, the European Parliamentary report quotes Ould Laghdaf as saying that 'we are the Maginot Line of the EU, which must help deal with this situation' (Panzeri, 2010: 10).[7] Reminiscent of coded threats made by Gadaffi to European officials (Tsourapas, 2017: 2377), this is a hyperbolic yet clearly calculated statement, given its appeal to a historically rooted European identity and its underlying crisis triggers. The fisheries agreement, meanwhile, he described as 'relatively balanced, given the balance of power' (Panzeri, 2010: 13), which might appear a coy, backhanded acknowledgement of the unparalleled importance of this agreement to the EU.[8]

While this modifies the balance of power between peripheral states such as Mauritania and the EU, it does not fundamentally alter the relationship of economic dependency between them. Indeed, the very resource at stake in this particular instance of migration diplomacy highlights the continued insight of Samir Amin's (1974a) theory of accumulation and development, which details how and why raw materials such as fisheries come to be exported from the periphery at a production

6. These were a return to constitutional rule and a cross-party national dialogue.

7. The Maginot Line was a line of defensive and offensive military structures installed in the 1930s along France's eastern border to protect against German invasion.

8. The EU-Mauritania Fisheries Partnership Agreement has been described within European Parliament library briefings as 'the EU's most important fisheries agreement both in terms of volume and product diversity' (Katsarova, 2013: 3).

cost lower than that available in the core. There is, in other words, no contradiction between the fact that migration has afforded increased leverage and agency to the Mauritanian state, on the one hand, and its peripheral economic status as a node for relatively low-cost extraction and export of raw materials, on the other.[9] On the contrary, the dispossession that has resulted from the opening up of the fisheries sector to foreign competition in West Africa has been a significant driver of the movements of people that the externalization process has been initiated to prevent (Cross, 2013).

In any case, as arrivals on the Canary Islands dissipated in 2007 and routes shifted northwards once more, the externalization process did not diminish in tandem. Instead, it expanded in form, shifting from an exclusively defensive and security-oriented approach toward a more preventative and holistic one. The idea, reminiscent of the ethos underpinning the Global Approach to Migration and Mobility (cf. Cardwell, 2013), has been that unwanted migration cannot be controlled through police and barbed wire alone. Instead, an approach more embracing of civil society and the perceived root causes of migration is needed. As we will see, however, the externalization process becomes beset by inertia and diverging interests in this preventative guise. In turn, this creates yet further opportunities for migration diplomacy and distinct state agendas to enter the fold.

The inertia of preventative externalization policies

Nestled in one of the older buildings that make up the grounds of the Mauritanian Ministry of Interior, there can be found an office with a sign on the door that says 'coordinator of the national migration strategy'. The coordinator in question is an ageing civil servant, described to me as 'semi-retired' by an EU technical expert who worked with him on implementing the migration strategy. On paper, the strategy is the primary means by which the externalization process was sustained in Mauritania after the crisis in Spain had subsided. But when we sat down to talk in his office, his top priority did not always appear to be ensuring the smooth implementation of EU diktats. Over multiple servings of tea, he railed against what he perceived as the arrogance and inflexibility of his European partners, lamenting the ingratitude displayed by the EU displayed towards Mauritania for the favour of agreeing to serve as 'gendarme of Europe'. As we have already seen, the Mauritanian state has derived distinct benefits from its adoption of this role, and so this sense of victimization is likely not so clear-cut. But it nevertheless illustrates the significant, mutually constructed, obstacles to the strategy's ambition of forging a comprehensive approach to migration management in Mauritania.

9. This point is nuanced somewhat in the next chapter, which shows from the vantage point of Nouadhibou how this peripheral status has transformed over time, and the role played by the EU's illegalization of migrant labour within this process.

The national migration strategy was born of a context in which the arrivals on the Canary Islands had, in its words, 'been brought under control' (Islamic Republic of Mauritania, 2010b: 5). For those actors that had to respond to the pressing problem that international migration now seemingly posed in Mauritania, opportunities arose from this juncture. How to sustain this response in a more sustainable and long-term fashion? How to embed the policy concern with migration that was sparked in 2006 more deeply within the Mauritanian state and civil society organizations? It was questions of this sort that an assessment bureau of French technical experts set up in Nouakchott in 2010 was tasked with addressing. For his part, the strategy coordinator lauded the level of professional dedication displayed by the team of technical experts. 'Every month,' he said, 'they came for ten days. And when they came, they worked! They worked 18 hours a day!' This work involved analysing relevant legislation and statistical data as well as carrying out interviews with public and private actors involved in 'the problematic of managing migration' (Islamic Republic of Mauritania, 2010b: 10). As such, this was less an exercise in approaching a pre-given field of actors invested in migration management, and more an act of summoning this field of relations into existence, sewing within it the discursive and financial means by which it could be reproduced (cf. Bacchi, 2009).

The end result was an 88-page document that provides a comprehensive overview of the migratory landscape in Mauritania and, on this basis, offers detailed recommendations on how it ought to be refashioned. In other words, the strategy identifies the main legislative and institutional instruments involved in the 'problematic' of managing migration and sets out the key axes along which this self-referential imperative can be carried out. This was both an empirical and a prescriptive exercise, in that it involved patching together existing statistical data from a range of sources while lamenting its paucity; outlining the existing 'legal arsenal' while critiquing its lacunae; and identifying relevant ministries, agencies, and civil society organization while also decrying the lack of coordination between them in matters of migration. On the basis of this sweeping analytical overview, the document offers a programme of action based upon four key 'strategic axes', namely: (1) the institutional framework for managing and measuring migration, (2) migration and development, (3) the promotion of the rights of migrants, refugees, and asylum seekers, and (4) the control of migration flows. Each strategic axis is further broken down into numerous projects, each of which has a dedicated budget, timeframe, and details of responsible ministries, agencies, and organizations (Table 2).

The net effect of such funding outlines is a financial hierarchy of recipients, within which government ministries and civil society organizations are differentially situated, and atop of which sits the EU. According to the strategy document, there were very few projects devoted to migration prior to the year 2005. Between then and 2010, external funds for migration-related projects reached €20.5 million, 77 per cent of which was financed by the EU, with a further 20 per cent being contributed by Spain, and 1 per cent by France (Islamic Republic of Mauritania, 2010b: 22). Of the thirty-nine projects supported by the strategy, six

Table 2 Breakdown of the national migration strategy's strategic axes and associated budget (Islamic Republic of Mauritania, 2010b)

Strategic axes	Amount	%
Institutional framework for measuring and managing migration	€3,790,000	31
Migration and development	€5,010,000	41
Promotion of fundamental rights of returnees, migrants, refugees and asylum seekers	€1,290,000	11
Controlling migration flows	€2,100,000	17
Total	**€12,190,000**	**100**

came under the responsibility of the Ministry for Foreign Affairs and Cooperation and eight under that of the Ministry of the Interior and for Decentralization. These two ministries take up an even larger share of the total €12,190,000 sum, absorbing €3,210,000 and €2,640,000, respectively (Islamic Republic of Mauritania, 2010b: 55). The diverse range of associations and bodies encompassed beneath the umbrella of 'civil society' had €2.5 million earmarked for them, with the view of enfolding them into the holistic approach to migration management envisioned within the strategy.

In this light, the national migration strategy appears an expression of what Mark Duffield (2010) once termed the 'liberal way of development'. When viewed as such, development is less about levelling out the global divide between rich and poor, and more about instilling within the latter an ethos of self-reliance, resilience, and making do with bare minimum basic needs. Given the challenge posed by undocumented migration to this divide, its policing and containment are a central element of the liberal way of development (Duffield, 2010: 62). For this reason, the migration strategy's expansion of externalization's remit from a defensive to a preventative logic is simply a means of diversifying and diffusing containment strategies. In practice, it has served this purpose by acting as what Gregory Feldman (2011a: 378) calls a 'mediating agent' within the border externalization process. In other words, it sustains relations between a diffuse set of actors, such as national ministers, EU experts, civil society organizations, journalists, etc., and orients them around a common framing of migration, namely the need for its containment through the liberal way of development. This has been concretely achieved through, for instance, a proposed inter-ministerial national committee bringing together eight different ministries headed by the prime minister charged with overseeing the projects financed under each of its four axes.

While this makes the national migration strategy a key driver of externalization in Mauritania, it has in practice been far from effective in achieving its goals. Indeed, of the €12.9 million earmarked within the national migration strategy, only €6 million has been spent, according to the coordinator of the national migration strategy. For him, the fact that EU financing takes the form of donations rather than loans means that their conditions 'are very, very difficult to satisfy'. In the eyes of an EU technical advisor who worked on the strategy's implementation, it was, in retrospect, 'too ambitious' (interview, Nouakchott, 26 March 2018). For this official, this inflated ambition was evidenced in a 2015 decision by the EU to close a Project Management Unit following a review of its activities. The Project Management Unit was set up under the strategy's first 'strategic axis', concerning the institutional framework for managing and measuring migration (see Table 2). Ultimately, however, the projects intended for management by this unit 'were unable to provide updates … were sometimes unable to even get started' (interview, Nouakchott, 26 March 2018). This is but one example of a broader trend of low project progress and completion rates, the consequence of which is that much of the funding promised on paper does not get released in practice: 'the biggest problem that we have within all of the development programmes in Mauritania is a low rate of [fund] disbursement' (interview, Nouakchott, 26 March 2018).

While the logic of the strategy is a diversification and diffusion of strategies for containing those on the move to Europe, its implementation in practice has thus been beset with inertia. The asymmetries of the Mediterranean centre-periphery boundary may manifest within both preventative and defensive guises of externalization, but this can just mean things grounding to a halt, as opposed to rapid deployments of naval interception missions. That said, not all elements of the migration strategy were equally fruitless. On the contrary, its unfolding has been highly uneven, with certain aspects seeing far greater degrees of progress and completion than others. What underpins these disparities in the strategy's implementation? And what do these gaps tell us about how externalization intervenes within the history detailed in the previous chapter?

Border externalization's colonial conditioning

Certain elements of the externalization process in Mauritania overcame the inertia discussed above to see significant progress, beyond project proposals and budget projections. What drives this uneven implementation? How to explain which axes and projects are driven forward with enthusiasm, while others are let fall by the wayside? To answer these questions, this section returns to the three dimensions of territorial organization that structured the analysis of the previous chapter. In this light, the colonial conditioning of the externalization process becomes apparent, as does the state's contentious relationship to this historic entanglement. That there is a general colonial conditioning of externalization is not in itself surprising. After all, as we saw above, externalization is an expression of the liberal way of development, a way of development which itself finds strong resonances with the era of liberal colonialism, and with the governance form of indirect rule in particular (Duffield and Hewitt, 2009; cf. Gaibazzi, 2023). But when situated within particular histories, such as that detailed in the previous chapter, the form and contours of these resonances become more precisely apparent. As we will see, decisions made in the colonial era about borders, mobility, and belonging continue to undergird similar decisions today, even when international experts and state officials disagree over the details. Indeed, as we progress through each of these three dimensions of territorial organization, the locus of agency progressively shifts from EU officials to the Mauritanian state, with some consternation on the part of the former about this dynamic. At the same time, however, the conditions in which this state agency is exercised remain those inaugurated by the colonial project.

Bolstering the linear form of territorial division

The headquarters of the IOM in Nouakchott lie within a walled-off compound on a quiet street in the neighbourhood of Tevregh Zeina. Queries can be made through an elevated window in the wall, where a staff member is stationed to field the questions and concerns which invariably come their way throughout the day. Plastered on its high walls, somewhere below the barbed wire, are posters with information about the IOM's 'assisted voluntary return' programme. Much like European territory, it is a lot easier for most to receive information about how to return to one's country of origin than to physically enter the interior of the compound. Like the offices of many such intergovernmental and nongovernmental organizations in Nouakchott, the IOM office is an expression of a more general bunkering of the aid industry, which increasingly walls itself off from the societies in which it intervenes (Duffield, 2012; Smirl, 2015). The Nouakchott IOM office is, in this light, a fitting spatial corollary to the logic of containment embodied within the national migration strategy. Indeed, as one of the primary intermediaries between the EU and the Mauritanian state in the area of migration and borders, the IOM has been central to the uneven implementation of the migration strategy discussed above. And it has been especially central to one element of the strategy that has advanced perhaps further than any other, namely a project of upgrading and modernizing the borders drawn during the colonial era.

The border infrastructure upgrade project emerged from the migration strategy's 4th strategic axis, aimed at 'controlling migration flows'. The project envisioned providing Mauritanian authorities with 'the technical means necessary for better checks on flows of entries and exits on state territory while respecting international human rights conventions' (Islamic Republic of Mauritania, 2010b: 76). Following the measures taken against those attempting to reach the Canary Islands post-2006, the strategy document explains, the Mauritanian government decreed the creation of forty-five new mandatory entry points into the country, with a view to better managing its terrestrial borders. It was hoped that this would facilitate the 'management of foreigners, including those in an irregular situation', through the standardization of entry and exit data and systems across national territory. A range of actors and interests have coalesced around this project, availing of EU funding to give life to the state's newly decreed territorial entry/exit points. Some have been equipped from scratch, and others significantly revamped. Having been drawn up at the turn of the twentieth century by Xavier Coppolani, these territorial borders are now infused with unprecedented digital validity and linear precision as a result of the border infrastructure project.

The IOM has been the most prominent implementing partner agency in this project. As a result, according to a former head of mission in Mauritania, border management has been the organization's biggest area of intervention in the country. This intervention typically takes two forms. It can involve equipping border posts through the installation of the IOM's Personal Identification and Registration System (PIRS), which has been piloted in Nouakchott Airport and Rosso border crossing. Designed by the IOM for the purposes of modernizing border management practices in the developing world, PIRS is 'an entry-level border management technology designed to be a key first step towards computerized immigration processing' (Frowd, 2014: 10). Secondly, the IOM also runs capacity building and training and workshops for Mauritanian security forces tasked with patrolling these new entry/exit points. At such events, international experts armed with knowledge of best practice in the world of border management impart these skills and practices to their invariably less well-informed national counterparts. If the origins of these borders are rooted in colonial history, this unidirectional knowledge transfer in the present indicates how this history's power relations linger on in the present.

That said, the Mauritanian state has been a willing and enthusiastic partner in the border infrastructure project. Indeed, the Mauritanian government has taken the lead in the construction of five of the forty-eight new posts (interview with Director of Borders and Land Affairs official, Nouakchott, 11 July 2018). In such cases, the coordinating state body has been the *Direction Frontières et Affaires Foncières* (Directorate of Borders and Land Affairs) within the Ministry of Interior. When discussing these efforts with me, the director of this body was eager to emphasize that they also have a key role to play in the construction of all posts, beyond the five for which the state acted as lead operator. This is because the state ultimately holds the details and coordinates necessary to proceed with the construction of a border post: 'there are several factors that need to be taken into account in this regard. For example, is there an electricity network? Is there a water source nearby? Is it accessible? All of these aspects need to be considered before deciding upon the location of a border post' (interview with Director of Borders and Land Affairs official, Nouakchott, 11 July 2018). In addition to this control over determining the specific site of a given territorial limit, the data that is entered at the post once it has been constructed will, in theory, be run through the centralized node of the state police headquarters, the DGSN (*Direction Générale de Sûreté Nationale*). In practice, however, there are significant disparities in degrees of data centralization and levels of technological infrastructure between the forty-eight different posts.

So, while the border form of territorial organization is a colonial handover, there is now a convergence of state and EU interest around its maintenance. For the EU, the process of upgrading border infrastructure addresses security concerns around migration, while also being an opportunity to place itself at the helm of societal progress and development. As Philippe Frowd has observed:

> Each new post opening is marked by a press release from the local EEAS delegation and attended by an official from the delegation, the IOM head of

> mission and local security actors. Pictures are taken, and this photographic aspect of the project is central to the day-to-day routines of EU staff: before-and-after images of the border posts line the walls of the local EEAS delegation, alongside posters for other border control projects
>
> (Frowd, 2014: 232)

The explicit visual association between the transition depicted in the 'before and after' photos and the presence of the EU thus reifies a hitherto imaginary teleology, endowing it with material form. As such, they provide material affirmation of the conceits of what Samir Amin (1988) termed the Eurocentric ideological construct. Once absorbed, 'it becomes impossible to contemplate any other future for the world than its progressive Europeanization, which is simply the diffusion of a superior model'.[10]

The state, meanwhile, finds itself bolstered in the area of one of its core constituent domains, namely the capacity to control movement into and out of its territories (Torpey, 1998; Vigneswaran and Quirk, 2015). As much was said to me by a Ministry of Justice official when discussing the importance of the state within EU-Mauritania migration cooperation, which he did by emphasizing the need to balance 'the right to movement with the sovereign right of the state to create laws that regulate entry and exit' (13 August 2018). As we saw in the previous chapter, the bureaucratization of entry and exit regulations that occurred during the colonial era laid a foundation for this 'sovereign right' in the Mauritanian context. But throughout much of the independence period, it lay more or less dormant. The border spectacle on the Canary Islands provided an initial impetus for its reinvigoration, but it continued to be embraced in Mauritania afterwards, even as this spectacle faded from view in Spain.

The illegalization of migration

As arrivals on the Canary Islands diminished after 2008, primary responsibility for migration policing gradually shifted from the international level to the Mauritanian state. The nature of this transition is encapsulated in the fate of *Guantanamito* in Nouadhibou. As we have seen, the detention centre in the city served as a means by which illegalized migrants who had been intercepted at sea or arrested on land could be processed in Mauritania, before being further dispersed across West Africa. But reports of abuses at the centre brought it sharp criticism from human rights groups (Amnesty International, 2008: 22; Migreurop, 2010: 26), as well as from the European Parliament delegation mission discussed above (Panzeri, 2010: 11). The national migration strategy document acknowledges these concerns, calling for a dialogue between diverse stakeholders so that 'a consensual solution can be put in place concerning the detention centre in Nouadhibou' (Islamic Republic

10. Of course, this conceit comes apart in the process of everyday life at many of these border posts, as we will see in Chapter 6.

of Mauritania, 2010b: 76). At the same time, the strategy criticizes Mauritania's legal framework for managing migration, noting that 'the deterrent aspect of this framework is lacking' (Islamic Republic of Mauritania, 2010b: 19). It also laments the 'existing statistical production apparatus, which is totally deficient' (Islamic Republic of Mauritania, 2010b: 56). The ideal solution, therefore, would manage and deter migrants in Mauritania, while also aiding in the production of statistical knowledge.

Just such a solution emerged in 2012, the year *Guantanamito* was closed and a new biometric residence permit for foreign nationals was introduced in Mauritania. The residence permit was brought into administrative effect on 13 May 2012, through a circular issued by the Ministry of Foreign Affairs to diplomatic agencies, international bodies, and community associations (Boulama et al., 2017: 17). While an initiative of the state, it sustained the prerogatives of externalization, as was made evident in the circular note explaining that the new residence permit fell within the framework of 'controlling and managing migration flows' (CRIDEM, 2012). During the opening months of the rolling out of the residence permit, the only place in the country in which an application could be deposed was a civil registry centre in Nouakchott, meaning that anyone hopeful of obtaining one was obliged to travel to the capital (Traoré, 2013). This was one of several issues that arose during a series of consultations between state authorities and migrant community representatives that took place in Nouakchott in the months following the introduction of the permit. Other grievances included the requisite fee of 30,000 UM[11] and the cumbersome list of documentation required to apply, which effectively rendered the permit out of reach for the majority of migrants in Mauritania.

During the fifth and final consultation, held on 9 September 2012, the then Minister of Interior announced that migrant community grievances had been heard; the fee would be lifted and the required documentation would now consist solely of a valid national identity card (Al-Akhbar, 2012b). Heads of communities were informed that they had until 15 October to encourage their compatriots to register in one of the three national civil registry centres that had now sprung up in the country (one in Nouadhibou, two in Nouakchott). To sweeten the deal, applicants were told that they could expect their permit to be issued within seventy-two hours. People flocked to the civil registry centres, with a vast number of people registering over the coming two weeks. Scenes of chaos and violence were commonplace, with one Senegalese national losing his life in a stampede outside

11. This amounts to roughly €75. The Mauritanian ouguiya was revalued in January 2018, so that 20,000 UM would henceforth be worth 2,000 UM, 1,000 UM would be 100 UM, and so forth. Here and throughout the ethnographic chapters that follow, however, I use the old system when speaking of financial sums. This former system was very much still in vernacular use while I was on fieldwork, and by all accounts remains so at the time of writing.

a Nouakchott centre on 16 October (Al-Akhbar, 2012a). However, the seventy-two-hour timeframe came and went, and no permits were issued. Finally, on 13 January 2013, state authorities once more summoned community representatives. They were told the residence permits were now available to collect, but only upon payment of the original fee of 30,000 UM. When confronted with this about-face, the governor of Nouakchott responded that the previous declaration that the permit was free was an 'error in communication' (Al-Akhbar, 2013). In the view of one EU official however, there was no error in communication (interview, Nouakchott, 21 March 2018). This was simply the cheapest and most effective way for the state to obtain a statistical approximation of the number of foreign nationals residing on its territory – in accordance with the recommendations of the migration strategy – though without having to reciprocate by delivering residence permits. In any case, many within the migrant community now found themselves in the peculiar position of being statistically documented by the state while remaining physically undocumented themselves.

The outcome of this state manoeuvring was an effective illegalization of the residing foreign population, embedding the condition of deportability (De Genova, 2002) within the country's major urban centres. In the months following the January consultation, patrols and raids in migrant neighbourhoods and workplaces became a regular occurrence. Local media speak of people being arrested outside of restaurants, in migrant majority neighbourhoods, and at informal labour pick-up points (Thiam, 2016). A Malian community representative (interview, Nouakchott, 13 December 2017) described these times as follows: 'those days, they entered rooms and knocked at doors. They demanded! They were in front of the markets, the streets, the public spaces where you get taxis. Often they even stopped the taxis to ask for documents, to ask foreigners if they have the permit.' A vocabulary originally born of the earlier deportations in 2006 soon proliferated in the capital city. *Rafle* and *ramasse*, respectively meaning 'raid' and 'pick up', came to signify the police operations that now characterized life in many migrant neighbourhoods and workplaces. *Séjour*, meaning 'residence', has become a ubiquitous shorthand for the infamous residence permit, and is synonymous with the experience of interacting with state authorities tasked with 'managing migration flows'.

The relationship of these developments to EU border externalization is perfectly coherent, if not directly causal. For their part, human rights organizations and local media were quick to situate the new campaign of raiding and deporting would-be migrants within the context of EU security policies (Migreurop, 2010; Thiam, 2016; AMDH, 2019). Indeed, by enticing people to apply for a residence permit, the state acquired a numerical approximation of the number of foreign nationals residing in Mauritania, thereby addressing the statistical deficiencies identified in the national migration strategy. And in framing the subsequent roundups and deportations in terms of 'controlling irregular migration', the deterrent aspect of the institutional framework of migration, argued in the strategy to be 'lacking', could be said to have been strengthened. Initially introduced in 2006 and embodied in the 2010 migration strategy, the logic of containing undocumented movement outside of

Europe was now being implemented with enthusiasm by Mauritanian state actors, thanks to the residence permit.[12]

At the same time, however, the speed and ease with which a campaign of deporting dark-skinned foreigners took hold suggest something deeper to have been at play in the execution of this particular element of the strategy. Indeed, the biometric residence permit for foreign nationals was part of a broader state-led process, one that both revived older racial tensions in the country and strained relations with certain EU officials overseeing the externalization agenda.

The biometric renewal of racialized territorial belonging

In December 2009, the Mauritanian government issued an international tender calling for a 'biometric solution' to the question of ensuring the reliability and security of national documents (Portail Mauritanien du Développement, 2009). The contract was awarded to French biometric technology company Morpho, which would be responsible for supplying biometric services to a new national civil registry body, the *agence national pour le registre des populations et titres sécurisés* (ANRPTS) that was created by governmental decree on 27 June 2010 (Islamic Republic of Mauritania, 2010a). It was within the framework of the ANRPTS and the overhaul of the civil registry that the biometric residence permit for foreign nationals was introduced in 2012. Unlike in Senegal, where a biometrics project is being financed through the EU Trust Fund for Africa (Statewatch, 2021),[13] the creation of the ANRPTS was financed by the Mauritanian state to the tune of €18 million (interview with EU official, Nouakchott, 21 March 2018). Nonetheless, the ANRPTS plays a central role within the newly constituted field of migration management in the country. Indeed, central to its remit is the 'registering and updating identification data of citizens and foreign nationals residing in or in transit through Mauritania', according to the decree that created the ANRPTS (Islamic Republic of Mauritania, 2010a). A member of the agency's public relations team also underscored its utility to the externalization process when we met, emphasizing that their database has meant they 'have been able to assist the state in managing migration flows' (interview, Nouakchott, 2 August 2018).

While the ANRPTS thus fits within the logic and discourse of the externalization process, it also drove a wedge between actors invested in this process. Adam, an EU official who worked on the border infrastructure project, lambasted the ANRPTS and the difficulties it had created with the EU (interview, Nouakchott, 21 March 2018). In his view, the ANRPTS had simply taken the entry-exit database that an international team of experts had been working on for the border infrastructure project and used it for its own national documentation purposes. This national appropriation of an externally created database resulted in a temporary suspension

12. A similar episode unfolded in July 2022, albeit this time the residence permits were also delivered. This regularization campaign will be discussed further in the conclusion chapter.

13. We return to the EUTF in the final section of this chapter.

of cooperation with the Mauritanian government in the equipping of border posts: 'we continue to construct border posts, but we don't equip them with software or materials or anything. Because this agency says it's going take responsibility, take it! And us, we're not going to help' (interview, Nouakchott, 21 March 2018). By using a ready-made database for national purposes, the ANRPTS thus contested the uneven relations that characterize the project. As has been widely observed in scholarship on externalization, state actors charged with implementing the agenda of the EU are well capable of appropriating, shaping, and manipulating this agenda to their own ends (El Qadim, 2014; Berriane, de Haas and Natter, 2015; Cassarino, 2018).

The EU was not the only actor to be marginalized through the roll out of the ANRPTS. As noted above, the biometric residence permit rendered many within the foreign national community 'deportable.' But as a result of the legacies of the history detailed in the previous chapter, the parameters of biometric citizenship (Ajana, 2012) do not always align with those of national belonging in Mauritania. Indeed, the civil registry centres at which migrant communities gathered in 2012 in the hope of obtaining a residence permit were also sites of fierce contestation over what it means to be Mauritanian. Allegations of discriminatory practices at the centres became widespread, with claims that those of Afro-Mauritanian background were being singled out as not authentically Mauritanian, and therefore not entitled to documentation (Antil and Lesourd, 2012). A campaign group formed in 2012 to articulate these grievances, leading demonstrations which often led to road blockages, violent police dispersals, and street clashes. Calling itself *Touche Pas à Ma Nationalité* (Hands Off My Nationality), it argued that the biometric system was the latest in a long line of racist attempts to exclude Afro-Mauritanian groups from the national framework of citizenship and belonging. Phrases such as 'administrative deportation' (Jeune Afrique, 2011) and 'biometric genocide' (Dia, 2015) voice this sentiment.

Such phrases explicitly and consciously evoke the memory of 'the events' of 1989, but they also underscore the ongoing relevance of the colonial racialization of territorial belonging that underpinned them, explored in the previous chapter. As we saw, the conversion of the river into a line of territorial distinction was shortly followed by a decree deeming Afro-Mauritanians in the colony of Mauritania tax-resident in Senegal, effectively creating a novel link between racialized group belonging and colonial territory. This novel link engendered a racialized suspicion of the authenticity of Afro-Mauritanian identity which continues at ANRPTS civil registry centres and in society more broadly today. As we will see in later chapters, police in Mauritania are often accused of arresting and attempting to deport Afro-Mauritanian nationals, whom they suspect of being 'illegal' migrants. As mentioned in the previous chapter, what is unprecedented in the colonial encounter is not the *invention* of racial groups; the *biḍâni* imaginary had long positioned itself in opposition to other darker-skinned Africans before the colonizer arrived on the scene.[14] What is new is the assumption of an immutable association between a

14. Indeed, *biḍân*, the word designating the Arab-Moor class in the *ḥassânîya* dialect, is derived from the Arabic word, *abyadh*, meaning 'white.'

linearly demarcated expanse of territory and these pre-existing racialized groups. Only with this confluence can the deportation of those deemed not to belong become thinkable,[15] be they recently illegalized migrants or Afro-Mauritanian citizens.

While the racialization of territorial belonging that occurred in the French colonial era thus shaped the social impact of the ANRPTS, the proliferation of such biometric solutions also infuses the present with distinctly colonial relations between Europe and Mauritania (Grossi and Lepratti, 2021). The Mauritanian state may have financed the biometric rollout, even earning the ire of EU officials in the process, but Morpho, the French biometric solutions company that won the tender, is deeply intertwined with the EU military and security industry (Akkerman, 2021: 19). While the ANRPTS diverged from the desires of certain EU officials stationed in Mauritania, it thus also opened a new technological resource frontier to the European security industry. Much like with the example of fisheries discussed above, then, state agency can simultaneously break with EU expectations while reinforcing a broader centre-periphery economic structure between Europe and Mauritania, in this case through prising open new digital terrains of resource accumulation (cf. Grossi and Lepratti, 2021: 51).

As it becomes enmeshed in the social fabric in Mauritania, the externalization process is thus conditioned by the colonial restructuring of borders, mobility and belonging detailed in the previous chapter. In the process, the economic dependencies and Eurocentric ideological hierarchies mediated by the Mediterranean centre-periphery boundary are reinforced. That said, externalization's colonial conditioning has undergone significant shifts and convulsions since the mass arrivals to Europe that unfolded in the summer of 2015. In the final section of this chapter, we turn to the ramping up of externalization that has followed the mass arrivals to Europe in 2015, detailing from the vantage point of Mauritania certain convulsions and ruptures that have since afflicted both its material manifestation and its Eurocentric imaginary.

The Mediterranean, the Sahel, and the EU Trust Fund for Africa

The Mauritanian national migration strategy expired in 2018, but the financial rewards for migration control cooperation that it introduced have only increased since. The primary vehicle through which this has occurred has been the EU Emergency Trust Fund for Addressing the Root Causes of Irregular Migration and Displacement in Africa (EUTF), which was launched in response to the mass arrivals to Europe that occurred over the summer of 2015. Since then, externalization has been kicked into overdrive. In one sense, so too has its colonial conditioning and the broader logic of containment outlined above. Indeed,

15. Thinkable, not operable. As far as the latter is concerned, the machinations of the postcolonial state discussed in the previous chapter are of far more salience.

insofar as the EUTF consists of 'a paternalistic and educative design of power that underpins relations of trusteeship' (Duffield and Hewitt, 2009: 12), the connection between the development project and the era of liberal colonialism has been further entrenched. But at the same time, certain cracks and splinters in this design of power are apparent in the EUTF. In what follows, I detail two that become apparent from the vantage point of Mauritania. The first concerns shifts in the Eurocentric geographic imaginary of the EU that has emerged through the Trust Fund, which has reconstituted the Sahel as a peripheral zone of security risk and intervention (Andersson, 2019; Idrissa, 2021). Correspondingly, the Mediterranean centre-periphery boundary is, so to speak, receding further, having lost a great deal of its Eurocentric narrative force after becoming the world's deadliest border-zone post-2105. As we will see, this collapse in Eurocentric hegemony is not lost on those in the Sahelian periphery. This brings us to the second crack, which concerns externalization's bolstering of authoritarian politics and fascist impulses within and outside Europe, respectively.

Sahelian peripherality and the EU Trust Fund's shifting geographic imaginary

Across the road from the western walls of the French Embassy's sprawling compound in Nouakchott, there sits a smaller but no less securitized facility. Opposite its steel sliding gate entrance, under the shade of a tree, a young member of the Mauritanian security forces lounges on watch, an assault rifle slung over his shoulder. The premises over which he stands guard are those of the permanent secretariat of the G5 Sahel, a security cooperation framework consisting of five Sahelian states that was launched in 2014, as the French war on jihadism in neighbouring Mali began to take regional form. While the vigilance of even the most astute soldier likely fluctuates with heat levels throughout the day, there are enough reminders of violence and instability to keep them alert to passers-by. In March 2023, four jihadist militants escaped from a Nouakchott prison before being 'neutralized' in a military operation shortly afterwards, in an episode which harked back to a spate of jihadist violence that afflicted the country between 2005 and 2011 (Thurston, 2020; Tasamba, 2023).

In addition to hosting the task officers, auditors, and resources directorates that make up the administrative organs of the G5 Sahel, the permanent secretariat in Nouakchott is host to a support programme staffed by a small team of European technical experts (Civipol, 2019). The team is operated by Civipol, a French security services agency that serves as implementing partner to a range of EUTF projects across West Africa (cf. Stambøl and Jegen, 2022). When we met in the premises in 2018, our conversation quickly branched out beyond the issue of migration. Indeed, the G5 Sahel support programme illustrates just how widely the financial and thematic net devoted to tackling the root causes of 'irregular migration' has been cast since 2015.

Born largely of EU member states' inability to coordinate a joint response to the arrivals of 2015, the EUTF has a more flexible and ad hoc approach to project financing than other EU development aid instruments, with emphasis placed on adapting to needs as they arise (Lopez-Lucia, 2019: 24). It also has a broad, 'pseudo-causal' understanding of the root causes of migration (Zaun and Nantermoz, 2021), supporting projects that often have little on the surface to do with this objective (Lopez-Lucia, 2019). Valued at €1.8 billion,[16] such initiatives span development projects, migration and asylum policy, and the security arena (European Commission, 2021). In regard to the latter, the inclusion of a 'stabilization' mandate within the EUTF has anchored it firmly within the security and military domain (Raineri and Strazzari, 2019), facilitating military capacity building and training missions, such as the EU Training Mission in Mali and the EUCAP Sahel missions in Niger and in Mali (European External Action Service, 2022).

As a result, the value of project financing unleashed in Mauritania through the EUTF has increased markedly compared to that envisioned within the national migration strategy. Between 2017 and 2019 alone, a total of €137.5 million worth of Trust Fund projects were in progress in the country (Coopération: Union européenne – Mauritanie, 2018: 24; 2020: 20),[17] which is eleven times more than the sum earmarked for project financing over the entire eight years of the migration strategy's existence. The spectrum of projects funded ranges from, on the one hand, typical development interventions aimed at alleviating youth employment, fostering 'resilience' of vulnerable populations, or the construction of schools, to, on the other hand, hard-headed security initiatives, such as projects aimed at countering violent extremism, the development of border surveillance and rapid intervention teams, and the support programme for the G5 Sahel encountered above (see Table 3).

Thematic diversity notwithstanding, projects across the breadth of this spectrum all aspire towards an implantation of norms about what constitutes correct administrative procedure and best practice. In other words, they embody the hierarchical relations of tutelage that characterize liberal colonialism's 'paternalistic and educative design of power'. Included within the G5 Sahel support programme, for instance, is a regional security forces training programme, called the Sahelian Security College. The aim of the Security College, in the words of a G5 Sahel support programme official, is to 'train senior officers, the police, the gendarmerie, the National Guard and give legal guidance regarding border management'. The degree programme involves 'a week of exchange on corruption, a week of exchange on relations between security forces and citizens – what works, what doesn't work'. This state of being a source from which best practice

16. Of the €1.8 billion sum that it earmarks for the purpose of 'addressing the root causes of irregular migration', €1 billion is drawn from the reserve of the 11th EDF (European Union, 2016: 15).

17. The headline total value of projects listed in the 2020 *Coopération: Union européenne – Mauritanie* document is €84.4 million. However, the combined value of each of the projects it lists comes to €104.5. I am working with this latter figure in the above calculation and have included it in Table 3.

and know-how can be imparted to lesser skilled national actors, who are then equipped to apply it, is replicated across numerous Trust Fund projects, be they driven by a developmental or security logic.

This continuity with the paternalistic relations of tutelage of the colonial era is, moreover, clear in the minds of G5 Sahel support programme staff. When we met at the G5 Sahel permanent secretariat, one official reflected upon how times had changed since the days of early independence, when the director of police schools in the region was invariably from France. 'Today', he suggested, 'that's finished. We help in launching the tools, but the application has to be by local agents.' Crucially, however, 'the aim remains to transfer these tools. We don't plan to stay here forever'. There is, in other words, a continuity between direct colonial tutelage and postcolonial knowledge transfer, albeit one that everyone would rather be rid of. Strategic aims must continue to be dictated from outside, but ideally, and paradoxically, adopted by national security forces of their own volition. The national security forces, in other words, ought to be endowed with autonomy and ownership, while remaining in lockstep with EU security priorities in the region.

Awaiting such an ill-defined teleological horizon, the Sahelian state appears consigned to what Dipesh Chakrabarty once described as colonialism's 'waiting room of history': 'Indians or Africans were not yet civilized enough to rule themselves. Some historical time of development and civilization (colonial rule and education, to be precise) had to elapse before they could be considered prepared for such a task.' The presence of the G5 Sahel support programme suggests the remedial task of educating and shaping domestic agendas so that they may walk autonomously the path of civilization and development has survived well into independence.[18]

Contraction and collapse in the expanded Sahelian periphery

While striving to perpetuate a relationship of dominance and superiority born of the colonial era, the EUTF also shows signs of strain in this projection. EUTF funding is dispersed by means of a geographic parcelling of Africa, which is divided into three discrete areas of intervention: 'North Africa', 'the Sahel region and Lake Chad area', and 'the Horn of Africa'. Within this geographic imaginary, Mauritania sits within the 'Sahel region and Lake Chad area'. When set against the history of EU engagement in Mauritania, this indicates a further shift in the EU's conception of the geostrategic zone that it designates 'the Mediterranean'. This geostrategic zone was inaugurated with the 1995 Barcelona Process, and the

18. While professing to move this timeline forward, the support programme may have the precise opposite effect. As Eva Magdalena Stambøl and Leonie Jegen (2022) have shown, there are striking parallels between the business model of Civipol, the G5 Sahel support programme's implementing partner agency, and the colonial-era public-private partnerships through which policing and surveillance of French colonial territories were carried out.

Table 3 Breakdown of EU Trust Fund projects, agencies and budget in 2019 (Coopération : Union européenne – Mauritanie, 2020: 20)

Emergency Trust Fund projects in Mauritania in 2019	Amount (in millions)
Improvement of youth employability and capacities of SMEs in the development of the construction and public works sector with local materials and of professional training in school construction sites (ILO)	€3.2
Promopêche – creation of decent employment and consolidation of existing employment for youth and potential migrants in the artisanal fishing sector (ILO and GIZ)	€14
Promopêche – promotion of employment and improvemnet of living conditions of coastal artisanal fishermen, youth, and women on the surroundign areas of natural protected sites in north Mauritania	€10
Reiforcement of capacities for improvement migration management to protect child migrants from exploitation and trafficking	€3
Contribution to the reinforcement of migration and border governance and management and facilitating the return and sustainable reintegration of migrants in Mauritania	€8
Resilience for social and cultural cohesion in Mauritania	€6
Programme for the reinforcement of the resilience of urban and rural vulnerable communities in Mauritania (Safire)	€10
Support programme for the G5 Sahel – Phase I	€7
Support programme for the G5 Sahel – Phase II	€10
GAR-SI (*Groupes d'Action Rapides – Surveillance et Intervention au Sahel*)	€8.3
EU for the security-resilience-development nexus in Mauritania	€25
Total	**€104.5**

Euro-Mediterranean partnership agreements it launched (Cavatorta and Rivetti, 2014). While Mauritania was not originally included in these agreements, it has been long framed as a Mediterranean partner state as far as migration control cooperation is concerned. As mentioned previously, this occurred through the 5+5 dialogue, which was launched at the 1990 Western Mediterranean Forum with the aim of tightening security cooperation between five Mediterranean states north of the EU external border and the five Maghreb states (European Commission, 2022). Moreover, Mauritania was repositioned within the Barcelona Process's conception of the Mediterranean in 2008, when it expanded through the Union for the Mediterranean framework (Cardwell, 2011; Abbott, 2018). Now, however, in the Trust Fund's reframing of Mauritania as a Sahelian – as opposed to Mediterranean – state, there can be seen a subtle rearrangement within the EU's imaginary of the regions constituting its inside and outside, its centre and periphery.[19]

19. Of course, 'the Sahel' as a category of external intervention has its own history. See, for instance, Bonnecase and Brachet (2013); Mann (2021); Amselle (2022).

The subtlety of this shift does not detract from its significance. If the Mediterranean is a mediating boundary between a geopolitical centre and its periphery, it has since 1995 been an expansive one, which orders a gradated hierarchy of inclusion with reference to human rights and democracy promotion (Youngs, 2002). The Sahel, by contrast, represents within this imaginary an expanded periphery, in which security interests, as opposed to democracy or human rights, reign supreme (Stambøl, 2019; Lopez-Lucia, 2020; Idrissa, 2021). This rearrangement was foreshadowed in the Mediterranean's conversion into a 'macabre death-scape' post-2015 (De Genova, 2016), which has rendered the more lofty Eurocentric conceits increasingly unsustainable. The Eurocentric geographic imaginary embodied in the Mediterranean centre-periphery boundary is, in other words, receding, giving way to an expanded Sahelian periphery, in which externalization's logic of containment sheds much of its justificatory façade.

If recent events in the Sahel are anything to go by, however, the EU's liberal colonial design of power sketched above will have difficulty surviving this shift. At the time of writing, the G5 Sahel is verging on moribund, following Mali's withdrawal from the organization in June 2022. The context in which this withdrawal occurred was a series of military coups and unprecedented popular mobilization against Western powers in the region (Niang, 2022). As the former colonial power, France has received the lion's share of this popular ire, resulting most prominently in the withdrawal of its regional counterinsurgency initiative, Operation Barkhane. But the EU has not emerged unscathed, with a flagship military deployment in Mali also being forced to withdraw in May 2022 (Euractiv, 2022). While for G5 Sahel support programme staff in Nouakchott, there is a vague intention not to 'stay forever', departure dates are increasingly being unilaterally fixed by Sahelian states acting in accordance with popular demand. In such a context, any liberal colonial design of power that is premised upon nominal cooperation and shared agendas cannot reasonably be sustained. In both the receding Mediterranean centre-periphery boundary and the expanded Sahelian periphery, then, the Eurocentric imaginary that underpins the externalization process is disintegrating.

If the foregoing discussion sheds light on the shifts in the geographic imaginary that accompany the EU border regime's outward expansion, it tells us relatively little about what drives externalization in the first place. Why is it that the prime method of integrating immigration and asylum policies into EU institutional structures has been by relegating them to states outside of the EU? This is an important point in its own right, but it is marginal to the overarching analytical concerns of this book, namely the encounter between externalization and the people, places, and histories it meets. It does, however, require brief consideration for this purpose, for it holds insight into the internal contradictions at play within the faltering liberal colonial design of power under discussion here.

Externalization, between liberal delusion and fascist fantasy

At the conference in the Nouakchott Chamber of Commerce, a first long day of presentations was coming to a close. A representative from the European

Delegation was taking questions from the floor, having delivered a talk outlining the activities of the EU in the country. One young student, however, was more interested in discussing the mass loss of life at Europe's edges, accusing the EU of 'acting in its own interests'. The response from the EU representative was swift and unequivocal: 'of course the EU and its member states act in their interests. That means control.' Perhaps seeking to mitigate any dictatorial aura this rather blunt statement may have exuded, he then clarified: 'if we open the doors to everyone, then that will bring problems with populism, with the extreme right, etc.' For these reasons, he concluded in a stern yet measured tone, 'we have to manage, like the father of a family.' In one sense, this pitching of the EU as benevolent, mediating patriarch of the externalization process gives direct expression to the 'paternalistic and educative design of power' (Duffield and Hewitt, 2009: 12) that defines the connection between liberal colonialism and development. The choice of a paternal metaphor is, in this light, no coincidence; it shows just how resonant the liberal colonial impulse is with the EU's own self-image, and indeed its very sense of ontological security (cf. Haastrup et al., 2021).

At the same time, though, these words also reveal certain internal contradictions at work within this design of power. In particular, his evocation of the far right highlights the paradoxical relationship of the externalization process to the spectre of fascism in Europe. At once contradictory and constitutive, this relationship between fascism and externalization dates to some of the earliest attempts to outsource migration control policy at the EU level. As we saw in Chapter 1, the 2002 European council meeting in Seville took place in the context of far-right electoral gains in France and the Netherlands (Peers, 2006: 108). Immigration and asylum were accordingly placed at the top of the Council's agenda, in the hope that social grievances could be siphoned into the electorally safer domain of EU external relations. The result was an enshrinement of clauses on compulsory readmission and joint management of migration flows into all future association agreements with non-EU states (Lavenex, 2006: 342), making it a seminal moment in the integration of migration concerns within the domain of EU external relations.

If a key logic of externalization is thus an impulse to keep the far-right at the electoral margins in Europe, a perverse effect of its enactment is a growing fusion of EU policy and fascist tropes and fantasy beyond the external border. As the border regime reconstituted itself in the aftermath of the 'summer of migration' of 2015 (Hess and Kasparek, 2017), so too did the far-right Generation Identity movement, which saw in the arrivals of 2015 a validation of the infamous 'Great Replacement' conspiracy theory. Acting to defend Europe from its purported 'Islamization', the group chartered a vessel in the summer of 2017 to intercept migrant boats in the Central Mediterranean and return them to Libya (Walsh, 2017). In the process, they relayed a familiar narrative demonizing search and rescue NGOs as acting as a 'pull factor' for migrants departing the shores of North Africa (Wildman, 2017), and instead proposed working with the Libyan Coast Guard to ensure people were denied any possibility of reaching Europe.

While this stunt was short-lived, its objectives were upheld in a memorandum of understanding signed just three months later between Italy and Libya, which paved the way for the delivery of patrol vessels to the Libyan Coast Guard for the same purpose of intercepting and returning Europe-bound migrants (InfoMigrants, 2017). Meanwhile, throughout 2017, the EU had stepped up its operational support to Libya, with vast financial sums and equipment flowing to Libyan security forces via the EUTF (Directorate-General for Neighbourhood and Enlargement Negotiations, 2020). As departures were consequently suppressed, imprisonment, torture, and sexual violence rose to an industrial scale in Libya, leading certain German diplomats to characterize detention centres in the country as akin to concentration camps (The New Arab, 2017; cf. Heyden, 2022). While Generation Identity's media stunt was ultimately unsuccessful, the form and consequences of official migration cooperation with the Libyan state in the interim have surpassed even what the far-right would dare celebrate publicly. While Amin (1988: 114) highlighted an intellectual affinity between Nazism and Eurocentrism, then, it would appear this coalescence has since acquired further material substance.

For these reasons, the defence of the externalization agenda in terms of stemming the far right is not only politically noxious, but also rife with internal contradiction. As Xavier Ferrer-Gallardo and Lorenzo Gabrielli (2022: 650) argue, externalization's ongoing reproduction of the 'border spectacle' may represent 'a splendid gift to those who make their electoral business out of it'. This is one of two paradoxical 'externalities of externalization' to which they draw attention at the Morocco-Spanish border in Ceuta; the second is the bolstering of authoritarian politics within societies located beyond the EU external border. In this regard too, the Mauritanian state has been no exception. As we have seen, the elements of the national migration strategy that progressed most successfully were those pertaining to border management and security, and which strengthened national security forces more generally. For their part, EU officials in Mauritania lamented to me on numerous separate occasions the difficulties of working with what they described as 'a military democracy'. A former IOM head of mission, meanwhile, was more delicate in his portrayal of their relations with the state. Evoking 'the complexities of Mauritanian society', he observed that 'things take time here, and you need lots of patience and diplomatic skills and liaison skills, basically, in order to get things done' (interview, Nouakchott, 28 February 2018). Such statements reflect varying degrees of frustration with a perceived impingement of social norms upon governance ones, and a consequent divergence from the primacy of 'the rule of law' in partner states such as Mauritania. There is an irony to these frustrations, however, since, as Ferrer-Gallardo and Gabrielli (2022) suggest, so much of the structural appeal of externalization lies in its bolstering of those extra-legal practices which cannot be carried out quite so easily within EU territory, as can be seen, for instance, in the *Guantanamito* detention centre.

While rhetorically standing for the rule of law and democracy, externalization in practice thus reproduces authoritarian politics and mimics fascist fantasy, while at the same time justifying itself on grounds of keeping the far-right at the electoral margins. In this light, its sustenance of colonial-era power dynamics is not limited to 'the educative design of power' associated with the migration-development-security nexus embodied in the EUTF. Indeed, for Kwame Nkrumah (1965: xii), a key strategic rationale of the system of neo-colonialism consisted in 'exporting the social conflicts of capitalist countries', a rationale clearly at play in the words of the EU representative at the Chamber of Commerce in Nouakchott. The foregoing discussion will have shown the limits to what can be achieved even within the parameters of such a rationale. For his part, Nkrumah was well aware of these limits, suggesting that the neo-colonial system's strategy of exporting its social conflicts 'cannot endure' (Nkrumah, 1965: xii). These limits are laid bare in the present juncture, be it in the EU's metastasis of authoritarian politics and fascist fantasy, the contracting geographic imaginary of the Mediterranean centre-periphery boundary, or in the collapse of Eurocentric hegemony across the expanding Sahelian periphery.

Conclusion

The Mediterranean serves as a mediating boundary between the EU's conception of its inside and its outside. In doing so, it naturalizes a colonial assumption of superiority and inferiority that implicitly governs EU external relations with states south of the Mediterranean. The externalization process bears a contradictory relationship to this geographic imaginary and the colonial power dynamics that it holds as natural. For, on the one hand, these dynamics were given a new lease of life with the inauguration of the externalization process in Mauritania from 2006 onwards. In its hard, defensive guise, externalization clearly displays the political asymmetries mediated by the centre-periphery boundary, but new opportunities for leverage that the era of migration diplomacy has created are also apparent. With an eye on the economic dependency structures originally born of this boundary, however, these opportunities do not amount to an overhaul of the structural asymmetries of the centre-periphery dynamic. As arrivals on the Canary Islands diminished from 2007, externalization took on a 'softer' preventative form with the national migration strategy. Put differently, it diversified the means by which those potentially on the move to Europe are to be contained on the southern side of the Mediterranean centre-periphery boundary. In the absence of an emergency imperative, however, externalization becomes beset by inertia and divergent agendas. In probing those elements of the strategy that overcame this inertia, we saw how the colonial shift in territorial organization conditions the externalization process in Mauritania.

At the same time, however, the geographic imaginary of the Mediterranean has faltered in its Eurocentric duty of naturalizing European dominance. This can be seen in the reconfiguration of the EU's geographic imaginary heralded by the EUTF, which has seen the Mediterranean centre-periphery boundary receding while a newly constituted Sahelian periphery, consisting of naked security interests, expands. It is no coincidence that this expanded Sahelian periphery has been the site of such unprecedented opposition to European intervention structures in recent years; it is a material corollary to the normative collapse in European moral superiority that accompanies the Mediterranean's status as the world's deadliest border-zone. Meanwhile, a core driver of externalization inside Europe is a desire to keep the far right at the electoral margins, notwithstanding the perverse outcomes this impulse generates beyond the EU external border. If the geographic imaginary of the Mediterranean is infused with the self-righteous hubris by which Samir Amin defined Eurocentric thought and practice, the same cannot readily be said of the EU's contemporary metastasis of fascist fantasy and authoritarian politics.

Having been forged at a time of European ascent, the Mediterranean centre-periphery boundary is thus increasingly faltering in the face of the realities it meets and creates through the externalization process. If this is so, are there also epistemic openings that can be grasped? What forms of transcendence are to be found in the debris of Eurocentric dogma? It is with questions of this sort that the political limits encountered by the externalization process translate into analytical limits. There is only so far that an analysis can be taken if it perforce adopts the conceits of its object of inquiry, in this case the Eurocentric assumptions that inhere in the Mediterranean centre-periphery boundary. For as Meera Sabaratnam (2017: 50) observes, 'Eurocentric assumptions place limits on how the researcher can think about and engage with the politics of intervention and systematically excludes much of the story and the relevant actors.'

It is for this reason that she argues for integrating the perspective of the 'targets' of intervention, which is inherently more expansive than that of the intervener. For our purposes, this means looking beyond the exclusive domain of the actors tasked with implementing the externalization agenda, and towards those framed within this agenda as invariably on the move to Europe. Their absence thus far has been particularly conspicuous, since they are present within the very same urban spaces of intervention that have been detailed throughout this chapter. Everywhere outside the walls of the G5 Sahel, the European Delegation, the IOM office, and the French embassy in Nouakchott are segments of the social life that the EU border regime views as in need of containment, as we will see in Chapter 7. From this standpoint, we will also eventually see how the limits outlined above constitute epistemic opportunities.

Chapter 5

A Tale of Two Cities

Illegalized migration and regional mobility in Nouadhibou[1]

In a relatively plush hotel in the northern port city of Nouadhibou, delegates flocked to a day-long workshop for civil society organizations and local government officials working in the area of migration. The workshop aimed to present the findings of a recently completed mixed methods survey detailing the conditions of migrants in Mauritania. While not quite at the scale of the two-day conference in the Nouakchott Chamber of Commerce detailed in the previous chapter, the array of civil society actors and local government officials the workshop attracted nonetheless demonstrate the ongoing potency of migration as a field of social intervention in the city of Nouadhibou. During a coffee-break, I decided to pick up on a general theme that had so far underpinned the session, by posing the question of how long the 'problem' of migration had existed in the city. The response, as expected, was since the '*pirogue* crisis' of 2006. But foreign nationals had always resided here, have they not, I inquired. 'Oh yes,' my companion Mohammed Ould Ely[2] responded, 'my primary school teacher was from Burkina Faso.' While Nouadhibou has been a pole of regional migration since at least independence, these movements went ignored by security, humanitarian, and media actors. Instead, they formed an intrinsic aspect of the social relations and political economy of the city in the early decades of independence.

With the initiation of the externalization process in 2006, these regional dynamics that are embodied in the city would be interlaced with a new form of Europe-bound movement and an associated apparatus of migration management aimed at controlling it. As detailed in the previous chapter, this apparatus consists of an array of defensive and preventative policies, technologies, and agreements pertaining to migration and border management. As a prominent node of Europe-bound movement that originally prompted these externalization measures, Nouadhibou remains a strategic site of their enforcement today. As

1. A version of this article appeared in a special issue of *Environment and Planning D: Society and Space* (Ould Moctar, 2022a).

2. Not a pseudonym.

we will see, this takes not only the securitized form of land and sea patrols, but also a broader circle of actors tasked with the provision of humanitarian care to those contained in Nouadhibou. Indeed, 'within a matter of years', Jocelyn Streiff-Fénart (2011: 1) observes of the externalization process in the city, 'these mechanisms have produced an entirely new universe of people, practices, places, and technologies gravitating around "illegal' migration"'. How does this new universe interact with the pre-existing regional migratory profile of the city, embodied in people like Mohammed's teacher from Burkina Faso? And how has the economic development trajectory that shaped these older migration patterns interacted with the illegalization of migrant labour in the city that has occurred in more recent years?

This chapter addresses these questions by detailing how these twin elements of Nouadhibou's migratory profile interact with one another, in the process reshaping labour-capital relations and the political economy of the city more broadly. It does so by foregrounding the subjective experiences and aspirations of a range of migrants I met in the city, examining how these disparate subject positions relate to the border regime's attempts to control them. In this regard, the chapter is oriented by assumptions put forward by scholarship on the autonomy of migration. As we saw in Chapter 1, this perspective emerged in part as a corrective to fatalistic depictions of migration and border controls as vast behemoths, in which migrants are little more than hapless victims (Papadopoulos and Tsianos, 2013; Casas-Cortes, Cobarrubias and Pickles, 2015; Karakayali, 2018). From the autonomy of migration perspective, on the other hand, migrants are a constituent and at times transformative force in shaping the direction and content of migration and border control policies (Bojadžijev and Karakayalı, 2010; Mezzadra and Neilson, 2013; Ataç, Rygiel and Stierl, 2016). This chapter takes this emphasis on the subjective experience of mobility, as well as the living labour that enacts this mobility, in order to discern the relationship between the border regime's illegalization of migrant labour and the political economy of the city. From this vantage point, it becomes apparent how the decline in European dominance sketched in general terms in the previous chapter has played out in the context of Nouadhibou. As we will see, while the early periods of independence were marked by the exclusive economic dominance of Europe, capital is today more transnational in character. The illegalization of migrant labour in Nouadhibou through externalization occurs within this specific juncture.

The chapter develops this point by firstly turning attention to the relationship between illegalized Europe-bound migrants and the EU border regime in Nouadhibou, detailing three distinct variants that this relationship can take. Each is defined by a subjective orientation towards Europe on the part of the migrant, on the one hand, and the drive of the border regime to control and filter this orientation, on the other. But beyond this broad antagonism, each of these forms of encounter is marked by their own idiosyncrasies and ambivalences. The subjectivity of those on the move is, in other words, a driving element of the externalization process, but its sweeping diversity means that it cannot be grasped in isolation from its social and structural context. In the second half of the chapter, we take up this task by locating these experiences within the historical development and social relations of Nouadhibou's political economy. From this perspective, their subjective experience is situated within the arc of the city's political economy, which has transitioned from being an exclusive conduit for European capital and investment to a site of more heterogeneous trade and investment flows. As a result, the openings and constraints associated with the illegalization of migrant labour in the city come more clearly to the fore. Ultimately, however, we come up against limitations associated with the exclusive emphasis upon Europe-bound trajectories put forward within much critical migration scholarship.

Autonomous migration and the border regime in Nouadhibou

Nouadhibou occupies a unique position within the strategic purview of externalized migration governance. As both a point of sea departure on the Atlantic Route to the Canary Islands and the final Mauritanian city on the overland route to Morocco, the policing of migration here happens on two spatial fronts. Unlike other 'transit' migratory nodes such as Agadez (Arhin-Sam and Lambert, 2023), then, Nouadhibou is of both maritime and terrestrial strategic significance to the EU border regime. This significance is reflected in the fact that the city is the only portion of Mauritanian state territory to which foreign security forces are deployed for migration control purposes. As we will soon see, the Spanish Guardia Civil deployment in the city is a key means by which the border regime remains locked into the self-reproducing cycle by which the autonomy of migration perspective characterizes the struggle over the control of human mobility. In what follows, we detail three distinct forms that this interaction between the border regime and the migrants it seeks to control can take. Some are characterized by the antagonism and polarization by which the autonomy of migration often characterizes this interplay (Papadopoulos and Tsianos, 2013: 188; Scheel, 2013: 585; Karakayali, 2018: 315). Others express a more ambivalent subject position, defined neither by outright defiance nor by passive acquiescence to the diktats of the border regime. While thus varying in degrees of abjection and coercion, each of these three expressions of the encounter between the border regime and illegalized Europe-bound migrants cumulatively demonstrates a need to situate such dynamics more firmly within their social and structural context, a task to which we turn in the second half of the chapter.

Unstoppable forces and immovable objects

In a street in the neighbourhood of *Qairān* in central Nouadhibou, I sat on a porch chatting and drinking coffee with a young Mauritanian man named Ba. As he told me about leaving for Spain by *pirogue* from Nouadhibou in 2014, it became clear that beneath his gentle, soft-spoken demeanour there lay a steely determination.

His excursion turned out to be short-lived; after spending three months in a Red Cross camp in Spain, he was deported back to Mauritania. Despite having been foiled by various instruments of the border regime, and in particular the 2003 Mauritania–Spain readmission agreement, Ba was thinking about his next move. While these instruments of the border regime strive to ensure the external borderline remains fortified against unwanted migration flows, they have clearly not been much of a deterrent for Ba. 'Immigration is my virus,' he told me, 'it's like cigarettes.' Next time, he says, he will travel overland from Mauritania to Morocco and get a boat from there.

Ba's aspiration to travel overland to Morocco and to reach Europe from there points to the peculiar dynamics by which the EU border regime reproduces itself. As we saw in the previous chapter, the era of externalization dawned in Mauritania following a crackdown by Spanish and Moroccan security forces on migrants at the fences of Ceuta and Melilla in 2005, and a subsequent southward reorientation of Europe-bound migration flows (Goldschmidt, 2006; Choplin and Lombard, 2008). By the time Ba attempted to reach Spain in 2014, the defensive and preventative elements of the externalization process had been firmly implanted in Mauritania (Peregil, 2015), resulting in a much-vaunted 'closure' of the Atlantic Route to the Canary Islands (Casas-Cortes, Cobarrubias and Pickles, 2014; Directorate-General for External Policies, 2017: 17). More recently, however, sea arrivals in Spain again reached unprecedented levels over the course of the Covid-19 pandemic, resulting in a yet more aggressive externalization drive, manifesting in, for instance, renewed signing of bilateral migration control agreements between Spain and Mauritania, and in the push for 'Status Agreements' with Mauritania and Senegal which would create permanent deployments of Frontex staff to these countries (Statewatch, 2022; Strik, 2023).

This seemingly pointless back and forth between the border regime's efforts to control migration and the relentless desire of those on the move to proceed regardless can be elucidated through the autonomy of migration perspective. As Vassilis Tsianos and Serhat Karakayalı (2010: 377) suggest in relation to the question of the power imbalance between migrants and the border regime: 'the question is not who is the winner of this game, it is rather: who initiates the changes in its rules?' In this light, while there is a vast power asymmetry between illegalized migrants and the agents of the border regime, the latter nonetheless have to adapt to the unpredictable nature of their object of control. This is why the points of friction between the EU border and migrants hoping to cross it have shifted repeatedly between various departure points across Northwest Africa over the past thirteen years. It is for such reasons that Nicholas De Genova (2017a: 11) has suggested border regimes to be 'convulsive reaction formations, responding always to the primacy of the sheer autonomy of migration'. Indeed, the force of will necessarily underpinning this autonomy may be seen in Ba's likening of his desire to move with an addiction, something driving him forward in spite of himself.

Meanwhile, the Guardia Civil officers tasked with policing these unpredictable impulses and movements remain deployed in Nouadhibou for this purpose. Luis is one such officer. We met in the marble reception area of a seafront hotel in

Nouadhibou, a sea breeze wafting through the double doors to coat the interior in a thin quasi-permanent layer of sand, to which an Ivoirien member of the cleaning staff periodically attended as we spoke. In a trajectory that might be viewed as a metaphor for the ever-deeper encroachment of border controls into sovereign state territory, Luis was first stationed at sea off the coasts of Nouadhibou, but now patrols the streets of the city with Mauritanian counterparts. The deployment comprises twenty-five Guardia Civil officers, who cover maritime and land patrols, liaison work, and technical assistance. They also work with six designated Mauritanian officers. Five Spanish police officers, tasked with training their Mauritanian counterparts, also form part of the unit (Peregil, 2015). This extensive cooperation is framed by a sweeping range of bilateral Spain-Mauritania security agreements, partnerships, and memoranda of understanding (Casas-Cortes, Cobarrubias and Pickles, 2014; Guardia Civil, 2019). According to Luis, their role in Nouadhibou also involves gathering information about groups and associations involved in the area of migration in the city and reporting this to headquarters in Spain. While under the responsibility of the Ministry of Interior when patrolling Spanish territory, the Guardia Civil fall under the Ministry of Defence when deployed overseas.

Just as Ba's determination to reach Europe affirms the importance of migrants' subjective aspirations within the dynamics of the border regime, the Guardia Civil officers deployed in Nouadhibou also find ways of justifying its drive for control of mobility. While Luis empathized with people who have been forced by poverty and a lack of life chances to undertake journeys they never otherwise would, he maintained that 'you have to have limits'. The questions of where these limits are drawn, by whom, and whether they should indeed override other principles like national territorial sovereignty or freedom of movement, did not figure within these considerations. Luis did, however, lay emphasis upon the human rights component of their deployment, lamenting the difficulties involved in importing European rights standards to North African security partners.

This humanitarian framing of the Guardia Civil's role within externalized migration control operations is as old as these operations themselves. Indeed, the maritime interceptions off the Canary Islands in 2006 were soon draped in a humanitarian veil, which transformed the Guardia Civil in the public eye into the 'Guardian Angels of the high sea' (Andersson, 2014a: 139). While more window dressing than a fundamental policy orientation,[3] this humanitarian framing of migration management has created space for a less openly antagonistic social position than that which connects Ba to Guardia Civil officers such as Luis. This occurs through integration within the vast network of civil society organizations gravitating around the field of migration management in the city. Nonetheless, inclusion within the humanitarian sphere is premised upon the adoption of a particular perspective on illegalized migration, and it is subject to consistent

3. As Andersson argues, these operations were underpinned by a militarization of migration control.

scrutiny by those tasked with acting upon this perspective. As such, the tension between the border and the migrant remains intact in the humanitarian space of migration control in Nouadhibou.

The professional migrant and the humanitarian border

'What's your opinion about migration?' Having thus far spent the duration of my visit immersed in his phone, Amadou chose these words to address me directly. A subtle but unmistakable challenge lay beneath the question. Mohammed fell silent, while someone making tea at one end of the room continued pouring the liquid from one glass to the other as the pot whistled on a gas burner. Eager to seize the opportunity for dialogue, I shifted position on the worn-out mattress upon which I was seated. 'Freedom of movement', I said, returning his gaze, before stating why, in my view, the unjust disparity between my ability to come to West Africa at my leisure by virtue of being a European passport holder and the inability of people like him to travel in the opposite direction made it plain to me that freedom of movement is a universal value. To my surprise, however, he disagreed. To Amadou, freedom of movement simply meant the right to go and die in the sea. He described in vivid detail the traumatic experience of the journey in the *pirogue*: feeling the waves crashing down on top of you, reaching around to see if your companion was alive or dead, not knowing whether you are alive or dead. For Amadou, the 'phenomenon' of migration stems from people's lack of hope. But he has hope now. He has kids and a family. And he runs an NGO for migrants in the city. Now, he thinks we should devote efforts to stopping people from attempting the journey. Now, he says, he is opposed to migration.

Amadou can hardly be faulted in his desire to prevent others from undergoing his traumatic experience. The logic of the border regime, however, is such that this desire can be more readily accommodated than a more substantive conception of freedom of movement than that which he holds. Indeed, one peculiar side-effect of the foreclosure of legal avenues to Europe has been the production of circles of 'professional migrants', amongst whom knowledge of and experience in the clandestine migration circuit can serve as social capital amongst international journalists, NGO workers, activists, and researchers (Andersson, 2014a: 257). Given their concern with documenting and mitigating border violence, such groups often have quite a different stake in the border regime than the Guardia Civil. But they remain enmeshed within its paradoxical operations.[4] In this light, the dynamic interplay between Europe-bound migrants and the border regime takes on a different hue, becoming reconstituted in the form of the professional, civic-minded migrant, and what William Walters (2010) calls the humanitarian border.

The humanitarian border was an integral part of the externalization process as it unfolded in 2006. In Nouadhibou, this took the form of organized care provided

4. As, of course, do researchers.

by the Red Crescent to those detained in the *Guantanamito* detention centre that was opened in April 2006. As Paolo Novak (2022) has shown, the spatial dynamics of the location in which the humanitarian border operates shape its material form and how it is experienced. In this case, these spatial features were the cramped and unsanitary conditions of an abandoned school that was repurposed by Spanish soldiers (Amnesty International, 2008). As we have seen, this was but one of the externalization measures that followed from what Sergio Carrera (2007) argues to have been a manufactured political spectacle that emerged in Spain that summer. In this light, the enlisting of the Red Crescent to care for the needs of those detained in the centre was far from a simple disinterested gesture of care, but rather 'an instance where humanitarianism was being operationalized in an attempt to manage a political crisis' (Walters, 2010: 145). Nonetheless, Zein Abidine, a Red Crescent official who worked at the detention centre, was keen to underline their mandate of care when speaking to me (interview, Nouadhibou, 1 April 2018). It was the Red Crescent, he emphasized, which tended to 'the hygiene, food and health of immigrants during their stay'. But he conceded that the detention centre was run by the state, which oversaw the administration, processing, and deportation of residents, embodying what Polly Pallister-Wilkins (2015: 55) has described as 'the double-side nature of humanitarian governance'.

The relationship of humanitarian care to state control is, however, not a static one. As Jill Williams (2015) observes in her discussion of a shift towards 'contingent care' on the US-Mexico border in Arizona, the degree of emphasis can oscillate between these poles in accordance with the broader policy climate. My arrival in Nouadhibou in April 2018 appeared to come at the apex of one such pendulum swing, at least as far as the international scale of humanitarian action is concerned. The 'crisis' of 2006 had well and truly dissipated, the detention centre had accordingly been closed since 2012, and, with the spotlight of European media directed elsewhere, the humanitarian care imperative of the Red Crescent appeared relegated yet further backstage. In its place, there appeared what Ruben Andersson (2014a: 272) describes as 'a landscape of ruins, structures set up only a few years ago that are already falling into disuse'. Indeed, since the closure of what Zein Abidine called the 'welcome centre', other projects had dwindled as well (interview, Nouadhibou, 1 April 2018). At the moment, he said with a hint of reticence, their activities consisted of 'a first aid training programme funded by the ICRC (International Committee for the Red Cross), a health programme funded by the French, which hasn't taken place yet … and that's it'. In the words of another Mauritanian former staff member at the 'welcome centre', back in 2009, 'there was a huge team, but now, there's nothing' (interview, 12 April 2018).

While the humanitarian terrain has certainly evolved since 2009, the situation may be less a case of things evaporating into 'nothing', and more of certain responsibilities being shifted to different actors. Indeed, the relationship of humanitarian care to state control appeared in full swing during my interactions with *l'Organisation des migrants de Nouadhibou* (OMN). OMN, their representative Khadija explained to me, is an association representing twenty-four different communities whose nationals are based in Nouadhibou. While waiting for a meeting to start, she gave me a proud tour of the rather sizeable space that was their headquarters. The humanitarian orientation of the group was clear in the function of each of the rooms. One was a rudimentary bedroom for migrants who had either just arrived in Nouadhibou or who for other reasons were unable to find a place to stay. Others contained materials for the numerous skills training they offered – sewing, baking, and dying fabric. Another was a classroom for children of migrants who were unable to attend school due to not having the correct documents. These activities have afforded a level of access to the corridors of power, as was demonstrated by the illustrious array of photographs on the wall of their office. Here they were with the Ministry of Interior official responsible for the national migration strategy; here she was with a World Bank official; this was after a training workshop held by the IOM, and so on. Each photo further solidified the impression of OMN as a 'manifestation of the involvement of civil society in the general set of organizational devices aimed at migration management', which Poutignat and Streiff-Fénart (2010: 10) argue to have emerged in Mauritania via the externalization process.

Accompanied by OMN's evident access to the realm of officialdom, however, was the watchful eye of authority. Khadija emphasized their positive relationship to state authorities, saying that they work closely together, providing them with trimestral reports on their activities. The Guardia Civil are also invited to attend their monthly meetings, in the interest of maintaining 'good relations'. This might in one sense appear a win-win situation: good relations for migrant associations and ample intelligence-gathering opportunities for the Guardia Civil. But the constraints that come with the arrangement were hinted at by Luis when he reflected upon attending the meetings: 'it's really nice, to see all of these people from different countries getting together to discuss their problems. Of course, they can't do anything – their hands are tied – but it's interesting.' Becoming embedded within the 'set of organizational devices aimed at migration management' thus comes with a degree of scrutiny that perhaps would not otherwise be the case for certain enterprising migrants in Nouadhibou.

Beyond such trade-offs, this embeddedness refashions the dynamic between autonomous migration and the border regime, reconstituting it in the form of the professional, civic-minded migrant, and the humanitarian border. Clearly, this iteration is not as antagonistic as the first one we examined, respectively embodied in Ba and Luis. But there is little evidence of the excess of autonomous mobility over the tactics of bordering when it takes this form. Instead, establishing oneself as a provider of humanitarian care to destitute migrants is conditional upon adherence to the logic of control upon which the border regime rests, as evidenced in the acquiescence of OMN to the presence of the Guardia Civil at their meetings.

This humanitarian reconstitution of the migrant-border dialectic is a relatively new development in Nouadhibou, one which appears to have emerged from the 'landscape of ruins' that is the international humanitarian mandate of the Red Crescent in the city. From the perspective of those whose social standing largely rested upon this mandate, something has been lost in this transition. In the view of the former Red Crescent staff member, the 'social role' of migration management has been lost with the closure of the 'welcome centre'. Instead, he said, it was in the hands of the police, 'who care about their own interests'. The mandate of the police in Nouadhibou in the domains of migration and border management does indeed differ from that of the Red Crescent and the Guardia Civil. And it casts yet another light on the relationship between the border regime and the illegalized Europe-bound migrant.

Jurisdictional limits and ambivalent autonomy

It took a few days before Yousef, of Nouadhibou police immigration services, found the time to sit down to talk to me. Our first scheduled meeting never took place because he was busy 'overseeing a deportation', according to Nouadhibou police station officials that day. When we finally did meet, it soon became clear that the specificity of the police's role within the architecture of externalization largely hinged upon their monopoly over the act of deportation. 'The police', Yousef emphasized with a hint of pride, 'are the only body that have the right to deport, between the gendarmerie, the national army and the Guardia Civil' (interview, 6 April 2018). While the Guardia Civil occupy the unique position in Nouadhibou of being the only foreign security force deployed for migration control purposes, for Yousef, the police's legal monopoly over the act of deportation serves as safeguard to national territorial pride and integrity. More generally, as scholars of externalization have long argued, third-country national security forces and state officials are far from passive victims of EU bordering prerogatives; on the contrary, domestic security concerns and interests often align with such prerogatives (El Qadim, 2014; Cherti and Collyer, 2015; Cassarino, 2018; Frowd, 2022). The institutional validation Yousef seems to draw from externalization's mandate to detain and deport shows Nouadhibou police immigration services to be no exception in this regard.

Yousef's description of their activities at first appeared to corroborate the observations of the disgruntled former Red Crescent official: 'there is no social mandate', he said, 'it's just a question of verifying whether someone's documents are in order or not.' If the person is not 'in order', then deportation becomes the default response. In such cases, Yousef explained, there are two possible routes: one passing through Rosso, on the border with Senegal, and the other through Gogui, on the Malian border. Which of the two one gets pushed through depends on nationality and/or the post through which one has entered state territory. Upon reflection, however, he subsequently added there was a tendency 'to fall into the social'. This can happen because of the leeway he claims is afforded to migrants who do not have a residence permit but are in possession of a receipt indicating they are waiting for it to be issued. While the former Red Crescent official viewed the police as wholly inadequate to addressing the 'social' side of migration management, Yousef felt the fact of informal leeway and grace periods in the policing of the residence permit meant that they occupy a *de facto* social function. In either case, however, the assumed dichotomy between the 'social' side of migration management and its more hard-nosed security counterpart is, as the literature on humanitarian border shows, an expression of a single process of militarizing of migration management.

While all of this makes Nouadhibou police immigration services a key organizing institution within the EU border regime's deportation infrastructure in West Africa, its jurisdicational limitations also generate holes in this infrastructure. While an overarching objective of externalization is the diminishment of breaches of the EU external borderline, the legal reach of the national police of course stops at the Mauritanian territorial border. For this reason, despite the fact that many transit through Nouadhibou to enter the Moroccan-controlled Western Sahara on a northward journey to Europe, there is no reason for such departures from state territory to be prevented by Mauritanian authorities. Indeed, according to Yousef, virtually all of the one thousand or so monthly departures he estimated to take place from the city are via the land border with Moroccan-controlled Western Sahara, known as PK55. This is a particularly attractive point for those moving northward due to an abolition of visa controls between Morocco and numerous West African states. Moreover, Yousef suggested a similar number of people to arrive in Nouadhibou each month, demonstrating it to be a transitory space on a northward trajectory. Morocco has attempted to temper these entries into its territory by decreeing that nationals of multiple West African states entering Morocco for the first time do so by plane rather than overland. But as far as the Mauritanian authorities are concerned, according to Yousef, individuals exiting Nouadhibou overland are 'in order', and are therefore allowed to pass. At a regional scale, then, this is an example of the legal cracks and jurisdictional grey areas within the border regime that allows collective mobility to exceed the structures that regulate it (Tsianos and Karakayalı, 2010: 376).

At the level of the locale of Nouadhibou, however, there appears less scope for such collective excess to manifest. Camara, a young Guinean migrant worker I met in Nouadhibou, has Europe firmly in his sights, but is more patient than many other Europe-bound migrants in the city. Having already spent a year in Nouadhibou, he has surpassed the length of time spent by many of his compatriots here, most of whom had only arrived in the past few months. For those like Camara who have spent more than three months in Mauritania, obtaining a residence permit is a legal requirement. But this has proven next to impossible, he says. He is sceptical about the fate of the fee once handed over to the civil registry office: 'They take the money and keep it for themselves!', he said of the 30,000 UM fee residence permit fee. 'They always say "tomorrow", "tomorrow", but you never get it.' It is not difficult to imagine how the biometric residence permit could prove so elusive to migrant workers like Camara. In addition to a fee that equates to a month's salary for the average migrant worker in the informal economy, the list of requisite documentation consists of a valid form of identification, proof of address in Mauritania, a contract with an employer, and a judicial record issued by Mauritanian authorities. For many, being 'in order' is not as simple a matter as Yousef the police officer would appear to make out.

As a result, deportation is a real and present prospect for Camara and many other migrant workers in the city. Unlike the picture painted by Yousef, there appears to be little dialogue when the police van shows up. 'When they come,' Camara said, 'it's better to run.' According to Yousef, the police ensure that a representative of the deportee's country of origin is contacted, and that the individual is afforded the opportunity to pick up their belongings before their departure. When I asked migrant workers whether this was the case, the question was met with derision. One man from Mali recounted how the police entered his friend's home and stood behind him as he prayed in his bedroom. 'They waited until he had said "salaam alaykoum, salaam alaykoum",[5] and then they took him.' The ever-present possibility of such a brusque removal is a feature of the condition of deportability (De Genova, 2002), and it injects a degree of uncertainty into everyday life in Nouadhibou. 'When I'm walking down the street,' Camara said, 'I'm always looking over my shoulder.'

If the state of being 'in order' thus belongs to the realm of discursive fantasy as far as most migrant workers in Nouadhibou are concerned, their embodied experience can be equally far removed from the expectations of the autonomy of migration position. Camara, for instance, is determined to reach Europe through legal channels. He has a wife and kids at home who are depending on him, he says, and for this reason is taking his time and saving money until he can go legally. This aspiration diverges from that of Ba, and from the sensational depictions of young men scrambling from pirogues on the shores of the Canary Islands or scaling the fences of Ceuta and Melilla. Such images are common currency within the aesthetic circuit of the border regime, but they also resonate with an autonomist emphasis on migrants' contestation and defiance. Camara's patient desire to migrate legally can hardly be considered a direct contestation of the border regime, but nor is it a passive submission to it. Rather, it is a more ambivalent space in between.

Given the irreducible complexity of human experience, this space of subjective ambivalence is, in all likelihood, a common one. Amadou can be seen to occupy a similar position, with his trajectory from a previous daring and traumatic attempt to subvert the border regime to his current preoccupation with addressing the local grievances and conditions that push people to such desperate measures. There is no contradiction in such a trajectory. As Harsha Walia (2021: 15) observes, the freedom to move and the freedom to stay are intrinsically intertwined; underlying the suppression of both are interlocking structures of violence and crisis which are not of the making of those they affect. Ba, Amadou, and Camara may each represent distinct subjective orientations vis-à-vis the apparatus of externalization, with distinct elements of this apparatus, in turn, being tasked with managing these subjective aspirations. But each of these interactions between the border

5. The ritual of the Islamic prayer concludes with a double uttering of this conventional Arabic greeting.

regime and illegalized Europe-bound travellers unfolds in a broader structure of social and spatial relations. In the following section, we turn our attention more directly to this structural context, detailing how it shapes the relationship between Europe-bound migrants and the border regime in the city.

The political economy of Nouadhibou

Agency and subjective aspirations cannot be discounted in the analysis of the dynamic between the border regime and those it seeks to regulate, but nor can they be viewed in reductive isolation. Indeed, while Ba's likening of his desire to move with an addiction in one sense points to the incapacity of structures of control to contain subjective aspirations, it also suggests an element of compulsion to be at play within these aspirations. One particularly pertinent factor underpinning such a compulsion is the unforgiving socio-economic context in which he and many migrant workers in Nouadhibou fare. This section turns in more detail to this socio-economic context, examining how historical and contemporary features of the political economy of the city relate to the life histories and experiences of Ba, Camara, and other illegalized migrant workers in Nouadhibou. To do so, I take heed from other voices within the autonomy of migration literature, and particularly Sandro Mezzadra's (2011c: 137) call to place the illegalization of migration within 'a wider analytical framework that examines the transformations of contemporary capitalism from the point of view of living labour and its subjectivity'. What emerges from this perspective is, in certain respects, a familiar story: the early years of independence are marked by a tussle between neo-colonial economic projects and state-led development initiatives, which gradually give way to an era of structural adjustment, informalization, and neoliberalization. The latter phase is encapsulated in Nouadhibou's 2013 conversion into a special economic zone, and it is within this same place and time that the illegalization of migrant labour in the city becomes entrenched.

Regional migrations, dependent development, and neoliberal restructuring

On the porch in *Qairān*, Ba played with an infant as I sipped from one of the small plastic cups of *café touba* that I had purchased for us at a nearby street stall. The young child was the son of a woman who ran this stall, selling the reputed Senegalese clove coffee to her many compatriots who live in this area. From the early 1990s, the neighbourhood of *Qairān* has been gradually emptied of its local Mauritanian *biḍân* inhabitants, who had begun pushing the frontiers of the city northwards. Despite having physically left *Qairān*, some of these Mauritanian *biḍân* have retained a stake in the neighbourhood, by converting properties into rental accommodation for newly arrived migrants (Choplin and Lombard, 2008: 155). In the process, Nouadhibou's new status within the border regime has been grafted onto Mauritania's existing hierarchies, as wealthy *biḍân* draw rent from West African migrant workers living in what is often crowded accommodation, such as the *cité des Sénégalais* apartment block in *Qairān* (Choplin and Lombard, 2008: 155), and the *foyer Malien* from which the young Malian mentioned above had been living at the time of his alleged removal by Mauritanian police. At the same time, however, the presence of West African migrants has left its stamp upon the character of *Qairān*, which is today host to a range of West African restaurants, fabric shops, and money wiring services, as well as all manner of informal stalls, such as the one from which I had bought the *café touba*.

The sparse details of Ba's life history that he shared with me over coffee nonetheless sufficed to illuminate how the recent EU-impelled illegalization of migration fits within the city's deeper migratory history and political economy. Ba's father was a migrant worker from Senegal, who left his home during the early decades of Mauritanian independence and settled in the mining town of Zouérat, located 650 kilometres northeast of Nouadhibou. Here he married a local Mauritanian woman, and Ba was born soon afterwards. How Ba's father came to look for employment in Zouérat in the first place is a result of the establishment in 1952 of a multinational mining operation named MIFERMA (*mines de fer de Mauritanie*), seventeen years after the discovery of iron ore mineral deposits in the region by French colonial researchers. The majority of workers at MIFERMA were either Mauritanians or West African migrants such as Ba's father. Their European counterparts made up less than 20 per cent of the workforce, but earned nearly eighteen times more than the local 'subordinate staff' (Bonté, 1975: 98). This economic segregation had a physical corollary, as European staff were housed in what were luxury quarters compared with the slums in which the majority of the workforce resided. For good measure, these spatial distinctions of race and class were further reinforced by a wall separating the two staff compounds (Bonté, 1975: 98).

MIFERMA cemented an emergent relationship of economic dependence between Mauritania and the European metropole, while also garnering extensive social and spatial transformations within the country. Owned by a conglomerate of European steel companies in which France was the largest stakeholder, MIFERMA ensured a peripheral subordination of Mauritania within the global economy through a ramping up of iron ore exports over the course of the 1960s, while holding wages stagnant throughout this period of increased production (Amin, 1973). At the same time, a slew of regional infrastructural development initiatives emerged around MIFERMA's new export model. A railway was laid connecting Zouérat to Nouadhibou to transport the minerals extracted from the pits to the coastal city, where an ambitious port project was underway to facilitate their export to Europe (Bennoune, 1978: 42). French industrial interests go further back in Nouadhibou, which was earmarked for development by a French fishing company called *Société Industrielle de la Grande Pêche* in 1919, but the new port project expanded its industrial profile, while facilitating the scaling up of the city's fishing industry that occurred in the 1970s (Marfaing, 2005). At the moment of political independence, then, infrastructural development in Mauritania had been firmly harnessed to the interests of European capital.

While conditioning this postcolonial social and economic context, the dominance of European capital did not exclusively determine it. Indeed, as a Senegalese migrant in Mauritania, Ba's father embodied the deeply embedded regional dynamics of circulation and exchange that have long been a feature of what David Robinson (2000) describes as a 'Senegalo-Mauritanian space' (cf. Ould Cheikh, 2004). Nouadhibou, in turn, encapsulates the role that these regional migrations played within the city's urban infrastructure and economic output in the early decades of independence. At a time when a majority of the residing population in Mauritania were either nomadic pastoralists or sedentary farmers, the industrial development of Nouadhibou created a labour supply gap within a range of sectors, which migrant workers such as Ba's father helped to fill (Bensaâd, 2008; Ahmed Salem, 2011; Choplin and Lombard, 2013). While the MIFERMA mining operation represented a subordinated integration of the Mauritanian state within the world capitalist economy, it also saw for the first time the emergence of a multi-national urban working class in the country (Amin, 1973: 73). Heightened by extreme rates of exploitation and stark inequalities between West African workers and European bosses, this emergent class consciousness ended up posing a seismic challenge to the European steel interests that ran MIFERMA. Combined with a shift towards Arab nationalism on the part of the Mauritanian government in the early 1970s, a protracted and at times bloody labour dispute at MIFERMA led to the mining sector being nationalized in 1974. Moreover, the previous year Mauritania broke from the CFA, the colonial-era currency shared between fourteen West African states which solders significant aspects of their monetary systems to France (Samba Sylla, 2021). Ba's father's trajectory thus reflects a period in which there appeared a real and tangible prospect of political independence being translated into economic independence.

By the time Ba made his ill-fated adventure in 2014, however, the economic and migratory landscape of Mauritania had changed drastically. Indeed, the border regime's illegalization of migration in the mid-2000s came off the back of a period of protracted social crisis. As we saw in Chapter 3, Mauritania was beset by droughts in the late 1960s and 1970s which decimated rural nomadic livelihoods, swelling the urban margins of Nouakchott and Nouadhibou. In contrast to the organized labour and class consciousness of the MIFERMA workforce, axes of solidarity and resistance adopted by inhabitants of the *gazra* and *kebbe* more often fell along family, kin, and ethnic lines (Vium, 2016). This reflected a broader shift in this period away from class struggle towards various forms of collective group identity, and a concomitant informalization of urban labour (Ould Cheikh, 2014a). Ba's sphere of livelihood activities that he listed to me mirrored this precarity and informality. He counted amongst his professions football, barbering, mobile phone repairs in the market adjacent to *Qairān*, and dance. Camara too displayed a precarious versatility when I asked what he does for a living in Nouadhibou: 'Here you do everything!', he responded with a laugh, 'bricklaying, painting, tiling floors. You want a plumber? I'm a plumber!' This perennial precarity has moreover accompanied Camara from his home country, where he worked as a transport operator driving trucks: 'every two or three months, they fire you. You're always looking for a new job.'

While garnering precarity and instability on the lower social rungs, the state response to these transformations was often to mould domestic policy to a new international investment climate, in what Armelle Choplin and Jerome Lombard (2009) deem an economic strategy of 'extraversion'. First, as both internal and international migrants continued to arrive in Nouadhibou, the government sought to ramp up the city's industrial fishing capacity in response to these arrivals (Choplin and Lombard, 2008). While the fishing industry had long been a source of revenue and employment in Nouadhibou, it would from the late 1980s become the cornerstone of the city's industrial strategy (Marfaing, 2005). The first fisheries partnership agreement between the EU and Mauritania was accordingly signed in 1987, paving the way for the numerous other such partnerships since. Mauritania has as a result come to represent the most significant of the EU's third-country fishing partnership agreement signatories. Today, fishing vessels owned by China, Russia, and various EU member states pepper the sea horizon off Nouadhibou's coast.

Secondly, in acquiescence to the reigning neoliberal investment environment, Nouadhibou was in 2013 declared a special economic zone. The formal aim of the SEZ is to boost output in a number of sectors in the city through the enticement of foreign direct investment by means of tax exemptions and the alleviation of customs tariffs (GRDR, 2018). It has been hoped that the attraction of investment in the fisheries industry in particular would boost on-land processing, as opposed to the offshore freezing and processing for export with minimal employment or value-added scope, which has typically characterized the access granted to industrial fleets (UNCTAD, 2016). We will see shortly how this has played out as far as migrant workers in the city are concerned. Beyond these aims, however, the effect has been what Chris Holden (2017) has suggested to be a form of 'graduated sovereignty' which characterizes special economic zones in general. This graduated sovereignty is a result of the national territorial fragmentation that special economic zones engender. In the example of Nouadhibou, this territorial fragmentation is stated plainly in the text of the law that created the special economic zones: 'the free zone constitutes a particular customary territory, separate from the national customary territory, in which the customary regime defined by the present law is applied' (République Islamique de Mauritanie, 2013).

The neoliberal investment climate in Nouadhibou is not the only means by which this fragmentation occurs, however. As previously mentioned, Nouadhibou is the only portion of Mauritanian state territory to which foreign security forces such as Luis are deployed for migration control purposes. In other words, the policies aimed at creating a favourable environment for international investors unfold alongside the externalization measures aimed at controlling unwanted migration flows, detailed in the first section of this chapter. These two elements of Nouadhibou's exceptional status are not separate from one another; on the contrary, they pertain respectively to the conditions of capital and those of labour.

The border and accumulation regime in Nouadhibou

On the side of the road leading up to the pier of what is today the Autonomous Port of Nouadhibou, Camara sat with a group of his compatriots. Here, he and others illegalized by the border regime gather to look for work as day labourers. Workers generally tend to cluster according to nationality here, Camara explained. This way, if someone gets a job with an employer, he'll call on some of his compatriots to accompany him. This area of the port is colloquially referred to as SIJP. Named after the French colonial fishing company whose launch of operations spurred the first building infrastructure on the slender peninsula in 1908, it today designates a vast expanse of the southern end of the port, all of which is said to be owned by a wealthy member of the *biḍân* class. SIJP encompasses two factories, a large scrapyard, a medley of Romanesque colonial-era buildings that have fallen into disrepair, and numerous shops and boutiques, specializing mainly in fishing catch materials. What was then named *Port Étienne* remained a relatively small fishing village until the 1950s, when the MIFERMA mining operation in which Ba's father worked sparked the process of social and economic transformation outlined above.

While at that moment in time capital accumulation in Nouadhibou was firmly harnessed to European interests, today it is more transnational in character. According to Camara, many of the employers who stop by the labour pickup are Chinese nationals looking for labourers, in a recruitment style mirroring that which has emerged from Chinese investment in the construction sector in Zambia (Lee, 2014: 51). In the case of Nouadhibou, labourers are usually sought to work on industrial fishing trawlers and in the multiple fishmeal and fish oil factories in the city. As host to thirty-nine such factories in 2019, compared with eight in Senegal and three in the Gambia, Mauritania sits at the regional helm of this destructive new export sector (Greenpeace, 2019: 10). These factories are overwhelmingly concentrated in Nouadhibou, whose conversion into a special economic zone in 2013 likely lies behind the leap from eleven fishmeal and fish oil factories being registered in Mauritania in 2011 to twenty-nine in 2015 (Greenpeace, 2019: 12). Half of these factories were in 2016 reported to be Chinese owned (Baxter and Wenjin, 2016). Moreover, China was the biggest importer of Mauritanian fishmeal in 2019, followed by Turkey and the EU.[6] Itself a beneficiary of internal migrant labour within export-processing industries in its own designated special economic zones (Ngai, 2004), China thus now avails of a similarly profitable and exploitative model in Nouadhibou.

While this represents a significant globalization of trade relations since the days of exclusive European capitalist dominance, the border regime's illegalization of this migrant labour plays a key role in facilitating these transnational flows. At SIJP, the air is thick with the possibility of deportation. Camara and his co-workers described how the police van comes here and whisks people away on a regular

6. On the other hand, the EU was far and away the largest importer of fish oil from Mauritania that year (Greenpeace, 2021: 49).

basis. The net social effect is an effective subsidization of profits through the production of a vulnerable, exploitable, and – in this case – transient workforce, as has been widely observed in the literature on 'migrant illegality' (Calavita, 1998; Parreñas, 2001; Anderson, 2003; McNevin, 2006; Iskander, 2007). A human rights lawyer in Nouadhibou made this clear (interview, 8 April 2018) as he described a typical labour dispute between migrant workers and employers that he encounters in the city: 'they [the employer] say they'll pay someone 10,000 and then give 1,000. They say they'll give 30,000 and they pay 10,000.' In his experience, there is little recourse for migrants in such situations due to the fears of going to the police that undocumented status generates. The era of exclusive European economic dominance has long passed, but the border regime's illegalization of migrant labour in Nouadhibou ensures a favourable position for transnational capital remains.

Just how favourable a position this is can be seen in Ahmed's experience in Nouadhibou. Upon arrival in the city from a neighbouring West African state, he made his way to SIJP early in the morning. By that afternoon, he was packing and loading boxes into trucks in a Chinese-owned fishmeal factory. Here he spent three months, earning 2,000 UM per day (roughly €5). He described how fish was powdered, boxed up, and transported from one place to another in the factory. Migrant workers of diverse backgrounds were assigned to each of these stages of the process. Upon hearing that workers were given lunch on top of their wage in other factories, they decided to demand that they also be given either lunch or an additional 500 UM per day. The struggle fell flat, however, because some were reluctant to commit: 'If there was unity, they would have succeeded. But there wasn't any unity. That's the problem.' Even if unity had been forged amongst this particular group of workers, however, there remains a constant presence of newly arrived migrant workers at SIJP who are eager to make some quick money before continuing onwards without falling foul of the mixed brigades and joint patrols of the Guardia Civil and the national police. If border regimes offer capital a means of temporally regulating access to labour markets (Papadopoulos and Tsianos, 2013), the transience associated with the Nouadhibou crossing makes its migrant workforce of particular tractability vis-à-vis employers in the city.

Furthermore, this tractability is not confined to the factory floor and construction sites to which people are recruited from SIJP. Having spent his first few months in various factory and construction jobs, Ahmed now works in a language centre in Nouadhibou. This has provided a greater level of stability and increased his income, but his situation remains precarious. He has not yet been issued with a contract: 'They just say we'll pay you every month, there's no need for a contract.' He also had little success in obtaining the residence permit and voices a similar scepticism as Camara: 'You pay, and you'll get a receipt, but you'll never get the permit. Everyone knows this. Now, everyone knows you won't get the permit, so why pay?' While for Yousef, those who have made no proven effort to obtain the permit are inevitably of 'bad faith', for Ahmed, the idea of making an effort that is known in advance to be futile is a waste of time and money. In any case, he fears he would have little recourse should the school ever refuse to pay him due to the possibility that going to the police without having his documents in order would end in deportation. Having moved from a higher-paying job with a modicum more stability than that by which he was initially greeted, the potential for exploitation is maintained by the elusiveness of the residence permit.

If Nouadhibou's dual exceptionality consists, on the one hand, of favourable investment opportunities afforded by its special economic zone status and, on the other, of the unique enforcement of migration control by foreign security forces in the city, these twin elements of exception coalesce in the form of increased profit margins created by the depression of wages and conditions of migrant labour in the city. This occurs through the sense of deportability generated amongst migrant workers, which serves as an extra-legal means by which Nouadhibou's special economic zone can afford low labour costs to transnational capital. The autonomous subjectivity of migrant workers in the city cannot be viewed in isolation from this joint regime of externalized border governance and capital accumulation. Once viewed as such, however, it becomes equally necessary to factor subjective experience into the analysis of the border and accumulation regime, as I would discover soon after meeting Malata.

Splintering the border and accumulation regime

I first crossed paths with Malata outside Nouadhibou police headquarters, where he and a group of day labourers sat on the curb, their grubby work gear adding to a collective aura of dejection. He didn't have much time to talk at this particular moment, however; his construction team had been waiting in the hot sun while a dispute unfolded between him and their Mauritanian employer in the station. The employer was claiming to be unsatisfied with their work and refused to pay them. A team of six, they claimed they were owed 7,000 UM each for two days' work. They had come to the police to resolve the matter. Later on, sitting by a fruit stand in the *Qairān* neighbourhood, Malata told me with dignified triumph that they succeeded at the police station in getting remunerated. Unlike Ahmed and most of his other fellow migrant workers in Nouadhibou's informal economy, he didn't let his legal status deter him from taking a stand in such situations: 'yes, a lot of people are afraid, they think when you go to the police they will ask you about paper matters. But you're there to talk about job matters, not paper matters. I know my rights! A lot of people don't know their rights.'

Clearly, for Malata, the vulnerability and tractability associated with the condition of deportability are not an inevitability. This may be rooted in the fact that he has already been deported three times from Mauritania: once in 2016, and twice in the past ten months. He has the right to stay for three months without a visa, he says, and if he gets deported after that, he'll just come back again. Given Malata's willingness to stand his ground during the labour dispute, it would appear that the acuity of deportation's labour disciplinary effects wears off with repetition.

Of further relevance in this regard is the fact that Europe sits firmly within the realm of Malata's past experience. He told me of spending long periods in the 1980s and 1990s as a seasonal labourer on farms in southern Spain. He left Spain for the last time in 2005, just before the drama of the '*pirogue* crisis' in the Canary Islands gave birth to the era of externalization in West Africa. The migratory experience and ambition of many others encountered here have emerged in tandem with this process of externalization. As a result, the stakes of the journey are of a different order of magnitude than they were during the years Malata spent in Spain. This likely means that while being painful and traumatic, the experience of deportation from Mauritania does not carry for him the same air of finitude that it might for many of his co-workers in SIJP whose sights are set on Europe.

Of course, from the intrinsically Eurocentric perspective of the border regime, Malata's prior trajectory and subjective experience matter little. As Sebastian Cobarrubias (2019: 782) observes of the migratory 'routes-lines' that the ICMPD i-Map uses to trace migratory trajectories outside Europe: 'the routes are highlighted as an existing fact on the ground [and] the directionality of concern is Europe-bound.' In other words, the cartographic instruments of the border regime assume all manner of regional movements to be oriented towards Europe, in the process creating a 'spatial truth' which is acted upon by other actors within the border regime. In this light, Malata's arrest and deportation on three separate occasions by Mauritanian police tasked with policing migration are perfectly in keeping with the prerogative of suppressing potential Europe-bound movement. After all, each instance of deportation would have counted among the statistics presented by national security and Ministry of Interior officials at press conferences in which progress in the policing of 'irregular migration' is demonstrated. All of this means that the subjective position and experiences of migrants like Malata count for little in the formal calculus of externalization.

Nonetheless, during the labour dispute described above, this position and experience momentarily splintered the hierarchical social relations that make up the political economy of the city, whose scales the border regime has otherwise tipped in favour of capital. In not allowing himself to be intimated into the role of docile migrant worker, Malata refused the vulnerability and tractability associated with the illegalization of migrant labour. Of course, collective bargaining gains do not emerge exclusively from such individual proclivities. Much less do deeper structural transformations. But all the same, this should caution against an overly deterministic reading of how illegalization is experienced. After all, the literature on 'migrant illegality' shows it to be an inherently unstable category, whose blurred boundaries leave it open to contestation and resistance (Coutin, 2005; Iskander, 2007; Chauvin and Garcés-Mascareñas, 2012; Reeves, 2013). If the standpoint of the subjectivity of living labour allows a unique insight into the transformations of contemporary capitalism, it also safeguards against any foregone conclusions in the analysis of these transformations.

For those engaged in such analyses, the fact that Malata was not *en route* to Europe should give further pause for thought. As his example shows, the layer of social relations into which the border regime injects deportability is also a site in which this effect of externalization can be challenged and cast anew. This is the case irrespective of where Europe lies in the trajectory or aspirations of a given migrant worker. While the autonomy of migration perspective rightly suggests illegalized Europe-bound migration to be a transformative and revolutionary force, such an emphasis should not foreclose attention to collective experiences and dynamics in places far removed from the institutional 'centre' of the EU border regime, but in which externalization has brought its operations. If the border regime tends to project Eurocentric assumptions onto the aspirations of all migrant workers beyond the EU external border, this is all the more reason for critical analysis to refrain from doing so. As we will see in the next chapter, which takes us farther south to the Senegal River Valley border town of Rosso, this refusal to project Eurocentric assumptions onto the needs and aspirations of migrant workers creates openings to view the deeper colonial history in which the border regime intervenes. While manifestly apparent in Rosso, elements of this history also became clear during my time in Nouadhibou.

Conclusion: Excavating colonial history in Nouadhibou

In an elevated courtyard overlooking the labour-pick up site at SIJP, there lie some dilapidated living dwellings. Their rusted, corrugated rooves look as though they could be lifted clean off by a strong gust of wind. It was outside one such home that I met Brahim. He was sitting on his porch tending to six or seven goats that were penned into the yard in front of the house. He told me he rents this property from a wealthy member of the *biḍân* class: 'He has plenty of projects, so this isn't important to him,' Brahim said. 'The white Moors here, they have plenty of money *mashâ' allah.*' A hint of remorse seemed to dwell beneath this statement. He then began to speak of how 'white Moors' had set up large-scale agricultural projects in his home region on the Senegal River Valley over the course of the 1980s and 1990s. It was at this time, he said, that Mauritanians stopped understanding one another, in what I soon understood to be an allusion to the infamous 'events' of 1989–91, detailed in Chapter 3. 'Me who is speaking to you right now', he said, 'I have members of my family who have fled to Senegal and others to Mali. We don't know what became of them.'

Nonetheless, Brahim did not feel resentment towards those who orchestrated the expulsions. 'We have a saying,' he said. 'If a donkey kicks you, and you kick him back, then you're the same as each other. Which is to say that we who were killing each other back then, we're as bad as each other.' In what could have passed for a paraphrase of Marx's famous reflections on the relations between English and Irish proletarians (Marx, 1870),[7] Brahim then reflected: 'we must unite to develop. The rich try to divide us, they have us massacre each other, so that they can take what they want.' Brahim likely had in mind the inequality that

7. In a letter from Marx to Sigfrid Meyer and August Vogt, Marx observed how 'artificial' divisions between Irish migrant workers in England and the English working class were maintained by various instruments of the English ruling class, a strategy that was the secret to its rule (Marx, 1870; for further analysis of the relevance of these insights to contemporary migration and border regimes, see Cross, 2021).

has been exacerbated in his home region as a result of the land expropriations that followed 'the events', to which we return in the next chapter. But I couldn't help but feel his analysis of the social schisms that generate fertile conditions for capital accumulation could equally resonate at the SIJP labour pick-up not far from where we were standing.

Indeed, at such sites across Nouadhibou, the social effects of EU border externalization interact with a more general capital accumulation strategy in the city, in a fashion that casts a different light on the relationship between migrants and borders. As autonomy of migration scholarship shows us, the EU border regime is driven by an always unfinished struggle to control Europe-bound movement. While this means that the subjective aspirations and experience of those on the move to Europe are a driving motor of externalization, these aspirations and experiences are anything but monolithic. Ba and Camara, for instance, each embody the desire to reach Europe often attributed to migrants in the region. And while Amadou no longer holds this ambition personally, he remains active in the humanitarian circles that revolve around its management and control in others. Beyond this, however, their subject positions vary widely. As Mauritanian nationals, neither Ba nor Amadou is subject to the residence permit policing by which the 'control of irregular migration' is operationalized in Nouadhibou. But at different points in time, they each exemplified the figure of the Europe-bound *clandestin* that so preoccupies policy and media in Europe. Camara, on the other hand, diverges from this trope in his desire to reach Europe through legal channels. But as a foreign national in Nouadhibou, he comes up against the obstacles migrants face to regularizing their stay on Mauritanian soil, and for this reason remains policed as an 'irregular candidate for immigration' by Nouadhibou police immigration services.

All of this illustrated the vast diversity of experience and conditions encapsulated within migrant subjectivity. How to foreground the constitutive and transformative elements of this subjectivity in light of this heterogeneity? The remainder of the chapter took up this task by situating these disparate subject positions within the history and social relations of Nouadhibou's political economy. In this light, Ba's family history appeared interlaced with a postcolonial trajectory of neo-colonial dependent development, regional migrations, class struggle, informalization, and eventual neoliberalization. The conversion of Nouadhibou into a special economic zone in 2013 is a culmination of this latter trend, and it dovetails with the illegalization of migrant labour that unfolded in the city through the externalization process and was institutionalized with the residence permit. It is in relation to this combined regime of externalized border governance and capital accumulation that subjective aspirations and experience need to be located.

Doing so in this light has broadened the autonomy of migration perspective's emphasis on collective subjectivity, which has often been concerned with Europe-bound movement. As we have seen in Nouadhibou, the border regime facilitates the flow of transnational capital by illegalizing people where they are, irrespective of where they have been or are headed. It follows that those who have no intention of reaching Europe can challenge the effects of externalization where they are, as Malata's example shows. In the next chapter, we build upon this insight by tracing the experiences in Rosso of migrant workers for whom Europe is also a past experience. As we will see, doing so in this context opens a window into the border regime's imbrication within a colonial legacy of racialized expulsion and dispossession.

Chapter 6

'YOU CAN'T BE A MIGRANT AT HOME'

Returning the gaze from Rosso[1]

In *ḥassâniyya*, the Mauritanian dialect of Arabic, the Senegal River Valley town of Rosso is referred to as *lgwārib*. This is the plural of the word *gāreb*, meaning 'small fishing canoe' – what is in French called a *pirogue*. The name highlights the importance of the Senegal River and its daily crossings both to the material workings of the town and to how it is popularly imagined. On the other hand, within the discourse and aesthetics of the EU 'migration crisis' as it has played out on the West Mediterranean, the '*pirogue*' has acted as a watchword. Media depictions of the '*pirogue* phenomenon' evoke images of desperate people washing up on sun-soaked Spanish shores in rickety fishing boats (e.g. Mayault, 2017), thus ushering in the humanitarian imperative of care and control by which the response to the 'crisis' is typically characterized (Pallister-Wilkins, 2015; Perkowski, 2016; Andersson, 2017; Gazzotti, 2019). Crucially, as we saw in Chapters 4 and 5, this humanitarian response often legitimates and enhances border externalization as a strategy of EU migration management (Cuttitta, 2018). This strategy, in turn, sustains and depends upon the 'colonial erasure of previously existing polities and societies' (Cobarrubias, 2019: 782) that are necessarily built into the border regime's geographic imaginary of the regions beyond its external border.

Within this Eurocentric imaginary, Rosso is reduced to either a node to be traversed on an invariably Europe-bound northward journey or a point through which deportations aimed at diminishing the quantity of such journeys can be executed. A rather different conception lingers in the *ḥassânîya* name *lgwārib*, recognizing as it does the centrality of the crossings that connect social life on the opposing banks of the Senegal River. At the same time, however, the flows of goods and people that traverse the river and bind the two sides of the town are laterally bisected by a national territorial dividing line, which situates one half of Rosso in Senegal and the other half in Mauritania. A product of the colonial era, this riparian border is today fortified by the infrastructure of externalization.

1. An abridged version of this chapter appeared in an article for a special issue of *Geoforum* (Ould Moctar, 2022c).

Nonetheless, the socio-spatial dynamics that have emerged from the colonial conversion of the Senegal River into a territorial border act upon and transform the externalization process in this particular locale. In other words, the form the externalization process takes is deeply conditioned by the situated socio-spatial dynamics and historical legacies in the regions in which it unfolds. In the case of Rosso, these dynamics force compromises upon the infrastructure of externalization at the border, while also casting a different light on the migrant illegality that is produced by the border regime (Andersson, 2014a; Cobarrubias, 2019).

If, as argued in Chapter 4, border externalization is overlain by a centre-periphery geographic imaginary, then this chapter offers a critical perspective on the European centre from the periphery. In this sense, we depart from certain elements of Mezzadra's (2011c) gaze of autonomy which oriented the analysis of the previous chapter. As we saw, this provided a necessary opening frame for situating the illegality produced by the border regime within the social relations and historical trajectory of Nouadhibou's political economy. The reason it is at the same time an insufficient frame is acknowledged by Mezzadra in the same piece. By way of an opening caveat, he observes that this particular gaze of autonomy 'has been developed within research and political experiences based in what we might cautiously and in a provisional way call "European"' (Mezzadra, 2011c: 122). While not denying the necessity of a more global analytical viewpoint, he suggests the expansion of theoretical and empirical parameters required for such a viewpoint to be beyond the scope of his endeavour: '[E]ven taking into consideration central and eastern European specificities would require a significant refining and adjustment of the theoretical frame that will be developed in the following pages. Needless to say, this would be even more the case if additional histories of migration and capitalist development were to be fully considered' (Mezzadra, 2011c: 122).

As far as this book's analytical concerns go, this Eurocentric caveat manifested in the previous chapter in the form of an exclusive focus upon Europe-bound trajectories, at the expense of broader subjective aspirations and social relations in which they are embedded. This chapter overcomes this obstacle by moving away from Europe-bound movement, to incorporate residues of colonial history into the analysis of EU border externalization (cf. Brambilla, 2014; Lemberg-Pedersen, 2019; Stierl, 2019). More concretely, this means picking up the historical narrative of Chapter 3, which culminated in 'the events' of 1989, and exploring what became of some of those who were expelled from the lands of the Senegal River Valley over the course of the violence. As such, there is no global claim being made in this particular chapter. Rather, the aim is, as Chakrabarty (2001: 26) puts it, to 'return the gaze', by offering a viewpoint on externalization from the perspective of a particular place-specific and embodied setting (Novak, 2017), and its broader historical context.

The chapter proceeds in three stages. First, it ethnographically explores the border infrastructure in Rosso, detailing the formal procedures enacted there at the behest of international migration management protocol, as well as the informal practices of bribery and extortion that this protocol generates in practices. It then delves into experiences of migrant workers in Rosso whose prior trajectories are marked by the violence of the EU border regime. Many of the markers of the migrant illegality which are experienced viscerally and brutally elsewhere in North Africa are shown to fade from view in Rosso. At the same time, however, this very absence brings into sharper relief the social conditions of one of the 'previously existing polities and societies' that Cobarrubias (2019: 782) argues above to be erased within the geographic imaginary of the EU border regime. In this case, these conditions are those of individuals who have returned to Mauritania after being deported during 'the events' of 1989. The collective subjectivity that emerges from this viewpoint encapsulates within it what might otherwise appear to be distinct structural processes of border externalization, racialized expulsions, and accumulation by dispossession. Finally, the chapter returns to the Rosso border, drawing an alternative perspective on it from the heterogeneous collective subject position that has emerged from this regional history.

Migration management within structural peripherality

Zooming in to the locale of Rosso means unveiling centre-periphery dynamics internal to Mauritanian state territory. Indeed, despite being a regional administrative centre, Rosso is in many ways a neglected backwater. A vibrant centre of exchange when it served as a transport node within the trade routes of the Waalo Kingdom, it has today retained a degree of political and administrative significance in its capacity as the capital of the Trarza region, and namesake of one of six departments in this region. But it has also been beset by underdevelopment in the postcolonial period. While Rosso is counted amongst the six principal cities in Mauritania, its 33,577 residents nonetheless in 2013 made up only 2.7 per cent of the collective population of these six cities, with Nouakchott taking up 76.6 per cent (Islamic Republic of Mauritania, 2015: 19). The uneven relationship reflected in this demographic discrepancy between Nouakchott and Rosso is further evidenced in the infrastructural relationship between the two cities. As one travels south from Nouakchott towards Rosso, the road steadily deteriorates in quality, at times degenerating into nothing more than a dirt track, a physical representation of the neglect of the municipality of Rosso in favour of the capital during the post-independence period (Kamara and Laborie, 1988). The Rosso road was supposed to have been refurbished in 2014 by an infrastructural development project financed by the tenth EDF but, according to EU documentation, the contracted company suffered from 'deficiencies' (Coopération : Union européenne – Mauritanie, 2018: 20). The project has been relaunched under the eleventh EDF, with €7 million being earmarked for its completion. In the meantime, the jarring nature of the journey south remains a fitting precursor to the experience of crossing the border itself, which usually entails negotiating with a bewildering array of individuals whose relationship to officialdom is often far from clear. As we will see, the presence of this extractive informal economy in Rosso is a symptom of its peripherality, which creates a zone in which the incentives and means for such practices flourish.

The infrastructure of externalization at the Rosso border

The border crossing from Mauritania to Senegal at Rosso takes place within a walled off area on the north bank of the Senegal River. In this area are located immigration and customs buildings, a ferry docking point, a shop, and a bank. Small canoes carrying people and goods land and unload regularly throughout the day. The ferry also crosses several times a day, albeit at indeterminate intervals, transporting cars and larger vehicles. For travellers willing to brave the rickety canoes and the negotiating tactics of their drivers, getting across the river can be done relatively quickly. For the less foolhardy, the ferry also lets individual passengers on board. This is nominally for free, but in practice every minute step of this border crossing is liable to be mediated by an informal charge. These water crossings may be rendered obsolete in the near future; a project to build a bridge over the river has been in the wings for several years but has yet to come to fruition. The EU has committed €20 million under the eleventh EDF to this endeavour, with the African Development Bank and European Investment Bank respectively pledging a further €40 million and €26 million. While works were due to commence in 2018 (Coopération : Union européenne – Mauritanie, 2018: 20), the project was formally launched with much fanfare in December 2021.

At the time of my visit in July 2018, Hamid, a customs officer, showed me around the border area. At the west end is a bank, a shop, and a gate through which larger vehicles such as trucks carrying goods long distance and camper vans pass upon completing the necessary paperwork. In the centre of the border area, in front of the ferry dock, is a larger building where entry/exit stamps are obtained, and visas issued for those obliged to obtain them. At the east end is the formal entry point into Mauritanian state territory for individual travellers. This is a single police building divided into four separate rooms, two by which those leaving Mauritanian national territory must pass, and two for those entering. Here police verify the documents of all those entering and exiting. Small-scale random custom checks can also occur at this point.[2]

2. Procedures differ for different travellers depending on the bilateral agreement in place between Mauritania and the country of origin. A vaccination certificate is required of all irrespective of nationality, though it may well be forged. Nationals of most states in the Global North are required to obtain a visa, as are those of many in the region, aside from Algeria, Libya, Ivory Coast, the Gambia, Niger, Mali, and Senegal. In the latter two cases, identification can take the form of a valid ID rather than a passport. In the case of Senegalese nationals in Mauritania and Mauritanians entering Senegal, fifty euro's worth of foreign exchange currency is also required as a deposit.

This space is a crucial node within the architecture of the EU border regime in West Africa. It is the national territorial endpoint of one of two principal deportation routes in Mauritania, the other leading from Nouakchott to Gogui, on the Malian border. While up-to-date figures are difficult to come by, Hamid claimed that buses deporting people from Nouakchott arrive here every day. A policewoman in Rosso Senegal also told me that they process deportees in the local police station every day. On the other hand, a member of the Red Crescent in Rosso Senegal claimed that the number was only a handful a month, a significant drop compared with the period from 2006 to 2010 (interview, 23 July 2018). This discrepancy could simply be a reflection of the number of deportees to which his association actively caters, as it was also clear that the roles of the Rosso police and the Red Crescent overlap: 'we're not police!' he eagerly insisted, before adding; 'but the commissaire here at the moment is good. He has the humanitarian spirit' (interview with Red Crescent official, Rosso Senegal, 23 July 2018). The affinity expressed here is likely a reflection of the fact that the police and the Red Cross in Rosso have each been contracted by Spain to perform complementary functions (Andersson, 2014a: 110). Whatever the exact numbers, then, the very presence here of the more general nexus of humanitarianism and border control, explored in previous chapters, is an indication of the Rosso border's instrumentality to the architecture of externalization in West Africa.

The border in Rosso has been cemented into this architecture by means of the national migration strategy's border infrastructure project discussed in Chapter 4. Hamid described how the architecture of the border had changed greatly over the twenty years that he worked here. 'The European Union did all of that,' he said as he showed me the police and customs buildings that were constructed as part of the infrastructure project. As Frowd (2014: 234) has observed, the form of these material structures corresponds to the particular way their financial benefactors (in this case, the EU) believe a modern border ought to function. This is observable in, for example, the institution of a clear spatial separation between entrance and exit, as the new police building leading out of the border area now ensures. The issuing of visas, entry/exit stamps, and other such documentation is also rendered distinct from their verification, through the confinement of each process to separate buildings, the former taking place in the building in the middle of the border and the latter in the police building at the east end. This creates an official division of labour and responsibilities between the two and – in theory – results in a smoother and more rigorous border crossing.

The border is further aligned with international migration and border management standards through the technological infrastructure that has been put in place here. As mentioned in Chapter 4, the Rosso border crossing is one of two territorial entry/exit points at which the IOM's Personal Identification and Registry System (PIRS) has been piloted (the other being Nouakchott Airport). This system serves to enshrine at the Rosso border 'two important technological trends: the use of biometric identifiers and the creation and integration of databases' (Frowd, 2014: 236). This combination of operational functions enables the connection of the data recorded at Rosso through PIRS to Interpol databases, thereby inserting it within regional and global security and policing networks. Adam, a former European delegation official based in Nouakchott, emphasized the utility of such data integration from the standpoint of wider regional security prerogatives when discussing an information sharing platform operated by the regional G5 Sahel force and supported by the EU: '[T]he idea is to share all of the information within the G5 Sahel countries. Even the number of migration flows. It's to come up with a bulletin that records everything: immigration, human trafficking, drug trafficking, arms trafficking, terrorism. A structured and complete database' (interview, 5 March 2018).

Thanks to the PIRS system, the Rosso border is thus integrated into a regional security apparatus whose logic groups migration beneath a broader umbrella of what are perceived and constructed as national security threats. There is also a utility to the state in this process, however, which Frowd (2014) illustrates through reference to James Scott's (1998) 'seeing like a state' thesis. From this perspective, such technological and physical infrastructure serves to enhance state legibility over its territorial and demographic purview. As Scott observes, however, this attempt to render the social legible inevitably fails to capture the informal social worlds onto which the state imposes its grid of intelligibility, a discrepancy that is manifestly apparent in Rosso.

Beneath the infrastructure of externalization

Beneath the more or less linear ordering of the physical queues at the Rosso border there lies an entirely different logic of access, one governed by informal social networks and relations. This vast network of unofficial relationships and practices is oriented exclusively towards extracting an illicit surplus from unsuspecting travellers. For the EU and the IOM, the national security forces stationed at Rosso function to ensure regional stability by managing the international flow of goods and people in an efficient and responsible manner (IOM, 2015, 2016). But for many police, customs officials, and intermediaries on the ground, the formal procedures impelled by migration management discourse simply become sites of informal revenue generation. This transition is encapsulated in the fact that informal charges and bribes are ubiquitously referred to as *les formalités* at border crossings such as Rosso.

The material infrastructure constructed by the EU contributes to this enterprise. In an observational report on the Rosso border crossing, Boulama *et al.* (2017) emphasize the long lines imposed upon travellers by the narrow pathway leading up to the police building. This makes conditions ripe for the emergence of micro-networks of patronage, given the universal demand for procedures to be expedited. The report describes this informal network as follows: 'administrative harassment is common currency, nourished by the security and defence forces (police, gendarmerie and customs officials) working with individuals who act as intermediaries for travellers. These intermediaries … take responsibility for administrative procedures alongside the security and defence force, for financial transactions carried out in total opacity' (Boulama *et al.*, 2017: 12). The density of these networks makes the passage extremely time-consuming and difficult, if not impossible, in the absence of an intermediary. Tourists, backpackers, and – alas – researchers are particularly vulnerable to this network, given how lucrative a 'Western' passport is in the eyes of intermediaries, and how patently out of place their bearers often appear in Rosso.

This informal network of extortion and bribery is a symptom of the peripheral condition in which Rosso finds itself, something suggested in a passing reference made by Hamid. 'We know everyone here', he bragged as he showed me around, 'the police, the guards, the *tcheb tchāba* running all over the place.' The *tcheb tchāb*[3] is something of a modern-day archetype in Mauritania. According to Zekeria Ahmed Salem (2001), the phrase evokes an individual able to make money by any means, usually through informal and perhaps illicit channels. It entails a strategy of 'getting by' through improvisation, of making ends meet in a spontaneous and cunning fashion. Akin to the figure of the grifter, in the Mauritanian context it is largely bereft of any negative connotation, being usually viewed as benign. While the *tcheb tchāb* can occupy any social rung, Salem argues the trope to have proliferated within the *gazra,* the informal urban settlements that have sprung up on the peripheries of Nouakchott over decades characterized by drought and a rapid, unwieldy urbanization. Christian Vium (2016) also highlights this connection in a discussion of '*Tcheb-tchib* strategies at the centre of the urban fringe in Nouakchott'. In the absence of any formal urban planning, *tcheb tchāb* strategies for acquiring, selling, and consolidating these illegally occupied plots of land have blossomed. Just as this urban peripherality has given rise to informal *tcheb tchāb* networks and strategies within the *gazra* shantytowns of Nouakchott, so too does the peripheral positioning of Rosso in relation to Nouakchott explain their ubiquity here. And it is, moreover, a reflection of a more general relationship between informality and peripherality (Larsont, 2002; Gilbert and De Jong, 2015; Oviedo Hernandez and Titheridge, 2016).

In Rosso, the glacial pace at which development projects such as the Rosso bridge and road construction projects proceed further compounds this peripherality and cements the zone of informality that emerges from it. Indeed, every person travelling to Nouakchott during the time I spent at the border was accompanied by an intermediary holding cash and papers. The costs may vary depending on extant relationships and whether or not the traveller needs a visa. But in any case, in the absence of an intermediary one is liable to spend an indefinite amount of time queuing just to obtain an entry or exit stamp. The €50 currency requirement for Senegalese nationals also creates long lines at the bank each day, where the requisite foreign exchange currency needs to be purchased, generating further means of extortion for *tcheb tchāba* intermediaries.

As a consequence, emotions and tensions unsurprisingly often run high at the Rosso border. In the brief time I spent there, European tourists being extorted by hustlers would occasionally lash out with emotive appeals to figures of authority,

3. *Tcheb tchāb* is the singular and *tcheb tchāba* the plural. The verb is *tcheb-tchib*.

whose reactions only exacerbated the frustration. A group of Senegalese nationals on a pilgrimage to a religious site in Mauritania also stormed the visa office and had to be forcibly removed, after their intermediary overstepped his boundaries during their *pirogue* crossing. Such scenes are commonplace, and they are a logical outcome of the insertion of international border management protocols and procedures within the space of informality that arises from structural peripherality.

For these reasons, while the Rosso border is a key technology of externalization, it is also a rather inefficient one. Cheikh, a Guinean migrant worker I met in Nouakchott, told me of how he made his way back to the capital a few days after being deported to Rosso Senegal: 'If you have money, you don't have a problem … The police down there are looking for money. They're men of the law but they do business as well.' Numerous testimonials gathered by Migreurop (2010: 31) further affirm this porosity of the Rosso border. In the words of one interviewee, 'even the police, they tell you "accept that you'll get deported and then just come back." The people who deport you, they see you two days later and say nothing.' The defensive logic of externalization thus necessarily exists alongside the informal economy at the Rosso border, and what Treasa Galvin (2015) describes as the 'normalcy' of deportation that it generates. While Rosso is the terminus of a state deportation route, then, being deported here is often far from the definitive endpoint intended by the logic of the border regime.

There is yet another more general aspect of social life in Rosso that poses problems for externalization's pristine infrastructure of state legibility installed at the border. This is the fact that the vast majority of border crossings into Mauritanian territory are undertaken by residents of the town who live on the Senegalese side and work on the Mauritanian side. Many are women who work as domestic workers or sell *bissam* and *tejamākht* drinks[4] or bags of rice on the street. 'They come here, every day, selling things, and then they go back,' Hamid explained, lazily observing the long line of people that snaked out of the entrance of the police building and around the path that leads up to it. From time to time, he would search people's belongings, verifying that there wasn't anything going by undeclared. While the vast majority of these crossings are for reasons of *petit commerce*, he explained that the goods they brought with them to sell were treated as 'personal items'. As such, they are exempt from formal customs charges, which of course is not to say that other charges do not materialize in practice.

These residents of Rosso making the daily crossing are exempt from the requirements imposed upon international travellers. Instead, Hamid explained, a *laissez-passer* card is required. Unlike the flashy biometrics of the IOM PIRS system, the *laissez-passer* is an entirely analogue technology, which has earned it the ire of some international experts, as we shall later see. It is a small card that can be obtained at the office of the mayor on either side of the town for a fee of 500 CFA and is valid for a month. Upon leaving their national territory, the cardholder leaves their ID card at the police building, presents the *laissez-passer* on the other

4. Local iced drinks made respectively from hibiscus leaves and baobab powder.

side of the river, and then recuperates the ID when they return. Some Hamid waved through without checking for the document. 'A lot of people who pass by every day I know to see, so I don't need to ask,' he said. The sphere of mobility created by the *laissez-passer* stops at each outer limit of Rosso, however, with police checkpoints at the limits of Rosso Senegal and Rosso Mauritania in place to enforce this delimitation. Here, the checks that are nominally situated at the border itself are effectively re-enacted in order to ensure that all those leaving the town hold the correct documentation and have carried out the requisite procedure at the border. The *laissez-passer* thus serves to preserve the town of Rosso as a contiguous sphere of mobility and circulation, notwithstanding the international border that bisects it. As such, it functions as a compromise between the colonial territorial division now sustained by externalization and the socio-spatial life bisected by this line.

In one sense, then, the Rosso border exemplifies aspects of the argument made in Chapters 3 and 4, which suggested the externalization process to be fundamentally conditioned by the colonial shift in territorial organization and the asymmetric centre-periphery dynamics that enabled it. The infrastructure of externalization in Rosso sustains the linear territorial function of the Senegal River that was instituted by the colonial encounter. At the same time, however, both the colonially imported border in Rosso and the infrastructure of externalization that emboldens it today have to reckon with concrete socio-spatial relations in this locale. In other words, the first pole of the border-migrant dialectic appears irrevocably conditioned by the nature of socio-spatial relations in Rosso. We now turn to look at the second aspect of this interplay – the illegalized migrant – to examine how this particular subjectivity interacts with these same relations and structures.

Migrant labour, dispossession, and displacement in Rosso

As we saw in the previous chapter, the border regime's illegalization of Europe-bound migrants in Nouadhibou, northern Mauritania, heightens the potential for labour exploitation within a diverse range of employment sectors in the city. This insight drew from existing scholarship that focuses on the precarities and violence experienced by those making northward journeys to Europe. But as Malata's example showed, the very fact of externalization means that EU bordering prerogatives reach people where they are, irrespective of where they are headed. We also saw how an exclusive focus on Europe-bound trajectories risks overlooking how deeper colonial legacies continue to shape migratory experience in the present. In what follows, we build upon this insight by examining the experiences of certain migrant workers in Rosso for whom Europe was a past experience, rather than a future aspiration. As such, it seeks to overcome the Eurocentric assumption that inheres in the border regime, which reduces all manner of circulations and experiences in West Africa to potential Europe-bound trajectories. Once rid of this perspective, the connections between the experience and structural position of migrant workers in Rosso and the regional history of racialized expulsions and dispossession detailed in Chapter 3 come into view.

The absence of migrant illegality in Rosso

One enters Mauritanian state territory via the police building at the east end of the border to come upon a narrow dusty road flanked on each side by stocky buildings. Most of these buildings are home to wholesale and retail shops, whose owners typically pass the hours lazing on sacks of rice observing the comings and goings of the entrance to the border, occasionally throwing themselves into the melee by barking orders and brandishing papers. Sandwiched between these storefronts and the flimsy rickshaws and large transport vehicles that often clog up the street, other less formal retail arrangements are in operation. Carts and handmade stalls that sell vegetable produce to those crossing the border throughout the day are also erected in front of the shops. It was here that I met Ali Bakar, who ran one of these stalls, and his companion Soro. Unlike the migrants who typically capture scholarly attention in Rosso, these men had neither been brought here from Nouakchott or Nouadhibou as part of a state deportation operation, nor had they just crossed the border on a determined northward journey. Nonetheless, the trajectories that led each of these men to Rosso are marked by the violence, exploitation, and vulnerability that are characteristic of the illegality produced by the EU border regime. Like the example of Malata in the previous chapter, these seemingly atypical trajectories and their situated experiences in Rosso offer unique insight into the externalization process.

Both men hail from Mali, Ali Bakar from a small village in the Timbuktu region, and Soro from the Mopti region. For each, quite a wide range of places and experiences stood between their initial departure from home and their presence in Rosso. Ali Bakar had previously spent six years working in a Turkish-owned factory in Libya, where his brother also worked. After the fall of Gadaffi, his brother made the Mediterranean crossing to Italy. According to Ali Bakar, he now works in a fish-processing factory in Malta. Ali Bakar also fled Libya, but in the opposite direction. He arrived in Mauritania in November 2017. After a brief stint in Nouakchott, he made his way to Rosso. Soro's trajectory was both more scattered and more violent. Having first left Mali in 2006, his narrative encompassed being intercepted at sea by Algerian security forces after leaving for Europe from Oran, getting robbed by Algerian 'bandits' in the desert, unsuccessfully attempting to enter Morocco from Algeria and being forced back by security forces and deported to Niger, and finally, participating on two separate occasions in an IOM voluntary return to Mali from Libya. Both Ali Bakar and Soro have thus experienced the physical, political, and biophysical violence (Squire, 2017) that characterizes the EU border regime's production of migrant illegality in the region (Andersson, 2014a).

In Rosso, however, many of the vulnerabilities associated with the condition of illegality were conspicuous by their absence. At his vegetable stall outside the border, within clear view of the police working the border post, Ali Bakar sold tomatoes, potatoes, onions, and carrots to cross-border travellers. At no point during any of the hours and days that I spent with him here did any interaction – hostile or otherwise – occur between Soro and Ali Bakar and the numerous police coming to and from the border. A similar peculiarity in the relationship to figures of authority emerged when Ali Bakar's landlord one day approached his stall. 'The money,' he demanded. 'Wait,' Ali Bakar answered dismissively, 'you'll get it later.' 'You always say "later", "later", you have to pay!,' the landlord responded. 'Later!' Ali Bakar repeated. The landlord paused for a moment, then: 'Give me a hundred ougiyas.' 'Why should I?' Ali Bakar responded, equally defiant. But the landlord's hand remained outstretched, until Ali Bakar reached in and gave him some change. The landlord took it and left without another word.

Both the surprisingly even footing of this exchange and the indifference of the police to the migrants' presence outside the border are reflective of a more general feature of social relations in Rosso. Unlike the cities of Nouadhibou and Nouakchott, there is a distinct lack of deportability as a lived experience of residents of the town. In the words of a member of the Rosso branch of the AMDH (interview, 14 July 2018), 'here there aren't deportations, in the sense of them going around the town, carrying out raids, and expelling people. I've never seen that. Never.' He juxtaposed the situation in Rosso with Nouadhibou and Nouakchott, where the infamous *rafles* and *ramasse* against which his organization campaigns in these cities are commonplace. He furthermore emphasized the distance between local authorities and the deportations that flow through Rosso from Nouadhibou and Nouakchott. 'We don't even see them here in Rosso,' he said of the deportees who get bussed in from farther afield. 'They go straight from the bus to the ferry' (interview, 14 July 2018).

This distance of local authorities from the practice of raids and deportations was further affirmed by Isaaka, a tradesman and representative of the Senegalese community in Rosso (interview, 23 July 2018). In his view, migrants in the town typically don't fear going to the police because they don't demand *séjour*,[5] as they are prone to do in Mauritania's other urban centres. Unless they have committed an offense or a crime, he said, 'migrants here go unnoticed.' This makes for an overall positive state of relations between the migrant community and local authorities: '[T]he judicial authorities here really help migrants. The law reigns.' The cumbersome documentation and prohibitive cost of the residence permit mean it is more than likely that neither Soro nor Ali Bakar is in possession of one. However, in Rosso this state of nominal illegality typically lies dormant due to the absence of its associated condition of deportability.

It would appear, then, that socio-spatial relations in Rosso do not just force compromises upon the border infrastructure of externalization and its colonial precursor. They also act upon the illegality produced elsewhere by the border regime, in such a way as to make it fade from the surface of lived subjective experience in this particular locale. Such absences are likely to be of little import to the border regime, whose gaze remains informed by 'pre-frontier detection' strategies aimed at pre-empting breaches of the external borderline (New

5. This shorthand for '*carte de séjour*' (residence permit) is regularly barked by Mauritanian security forces during raids on immigrant dwellings and workplaces in Nouakchott and Nouadhibou.

Keywords Collective, 2015: 75). But this glaring discrepancy between the content of the Eurocentric gaze of the border regime and social facts in Rosso underscores the importance of the latter to a non-Eurocentric perspective on externalization.

In what follows, I build towards such a perspective by examining what comes in the absence of illegality in Rosso. This involves taking what might appear to be an analytical detour, by briefly turning our attention to examine more directly certain aspects of social relations and structures in this particular locale. In practical terms, this means picking up the historical narrative of Chapter 3, which culminated in 'the events' of 1989, and exploring what became of some of those who were expelled from the lands of the Senegal River Valley over the course of the violence. As we will see, the aftermath of their expulsion and the demands of those who have returned together shed light on the structural processes that come to fill in for illegality in Rosso. In this way, the EU border regime's imbrication within this particular regional history comes to light.

In the absence of illegality: Biometric abandonment and racialized dispossession

Beyond the easternmost extremity of Rosso lies a small settlement that is home to around forty people. It consists mainly of a scattering of dilapidated tent-like structures made of tarpaulin and corrugated steel, with a handful of concrete buildings located in the centre of the settlement. Having initially been located just on Rosso's periphery, it was later moved farther east and now sits a few kilometres beyond the urban infrastructure of the town. There are 35 such settlements located in the Trarza region, with 118 in total dispersed across the Mauritanian side of the Senegal River Valley (interview with Rosso AMDH member, 14 July 2018). These are the homes of those who were deported from Mauritania during 'the events' of 1989 and have since been repatriated. While one might assume 'repatriation' to mean returning to settle anywhere on state territory, in practice the spatial scope of their return has been restricted to the five southern Mauritanian provinces of Trarza, Brakna, Gorgol, Guidimakha, and Assaba. A young Mauritanian Fulani man named Moussa showed me around the Rosso settlement.

Moussa was in secondary school when he and his family were deported from state territory in June 1989. He had grown up in Rosso and was living with his father. State security forces approached them in June of that year, he said, accusing his father of being from Senegal. He responded that 'he had no idea what Senegal looked like, that he had never been there'. Nonetheless, Moussa said, they were brought to the river, and forced to cross. He was one of tens of thousands of Afro-Mauritanians, the vast majority of them Fulani, deported alongside Senegalese nationals during the crisis. Upon arrival on the opposite bank, Moussa and his family were received by Senegalese officials and UNHCR workers, who registered them as refugees. He settled in a refugee camp in Dagana, one of many that had been established by the UNHCR along the south bank in order to accommodate the Mauritanian deported. While a minority of the displaced who were formerly members of the military and civil service were able to make strategic use of the

Figure 3 Moussa approaches his resettlement camp on the outskirts of Rosso. Photo by author.

refugee label in order to bolster their campaign cause, a majority fell into a zone of indistinction in which familial ties, fraudulent documentation and urban absorption all played a role in survival (Fresia, 2009: 348). For his part, Moussa never went back to school. Since he was the oldest of the family, and they were in need, he worked in rice and tomato fields in Dagana, picking at the time of harvest and sowing new seeds afterwards.

Moussa came back to his hometown of Rosso in 2008, thanks to a return programme implemented through a tripartite agreement between the UNHCR, Senegal and Mauritania. As we will shortly see, the experience of his community in Rosso since their return provides a window into the deeper historically constituted social relations in which border externalization intervenes. But the fact that he returned in the first place also points to the relevance of the broader framework of EU-Mauritania relations in which externalization is equally embedded. As mentioned in Chapter 4, the coup in August 2005 that ousted President Maaouya Ould Sid'Ahmed Taya resulted in a temporary suspension of aid flows from the EU to Mauritania. The EU provided the interim Military Council for Justice and Democracy with a sweeping range of undertakings, presented as prerequisites for the normalization of relations. Included amongst these was 'facilitating the return of refugees of proven Mauritanian origin and implementing any measures necessary for the reinstatement of their rights' (European Commission, 2006: 2). This exhortation paved the way for the creation of the return programme. By the time the August 2008 coup cut short the democratic reign of Sidi Ould Cheikh Abdellahi, the return programme had already been underway for eight months. Consisting in total of 49 'convoys', it transported 19,000–20,000 people across the Senegal River to Mauritania in trucks and canoes between 2008 and 2009 (UNHCR, 2009).[6]

Moussa came to Rosso with the 'third convoy' of returnees. This relatively early arrival meant that he and his family were offered housing in the buildings at the centre of the settlement. Later arrivals were accommodated in the rather more makeshift tarpaulin tents. This disparity in housing conditions is one manifestation of a more general sense of abandonment that has beset many of the returnees, which was expounded to me by Babacar, the head of the Rosso returnee community. He emphasized two elements of this abandonment. The first of these was the biometric civil registry system. In theory, the civil registry's recent biometric transition should not have posed a problem for the returnees, because they had already been registered by the UNHCR as part of the tripartite agreement and issued with a voluntary return form, entitling them to birth certificates and identity cards (Radio France Internationale, 2011). But, according to Babacar, 'that was before the arrival of the biometric system'. A protocol was put in place between the UNHCR and the Mauritanian state following the creation of the ANRPTS, in order to facilitate returnees' access to administrative documentation during the biometric transition. Civil registry centres were provided with specific procedures and registration equipment for the biometric registration of returnees, but many have since stopped functioning. In the case of the centre in Rosso, according to a member of the AFCF (interview, 17 July 2018), 'for two years now they've been saying that the machine isn't working'.

As we saw in Chapter 4, the biometric overhaul of the civil registry system is a crucial aspect of the context into which the EU border regime infuses the experience of migrant illegality in Mauritania. Despite having been born on Mauritanian soil, Afro-Mauritanians often find themselves deprived of access to biometric identity documents, and thus of suspected of being 'illegal' by Mauritanian security forces. Whether this is by accident or design depends on who one asks. In the view of a representative of the ADMH in Rosso, the exclusion of returnees from civil registry documentation is not an intentional policy, but an outcome of the difficulties of the biometric transition: 'No, I can't say it's done on purpose, because there was this change. In 2010, there was the new biometric system that was introduced.' In any case, the colonial racialization of territorial belonging detailed in the third chapter has been renewed with the biometric transition, in the process tying the experiences of migrants who have been recently illegalized as a result of EU bordering prerogatives to those of the racialized, displaced and dispossessed

6. While the '*pirogue* crisis' may have acted as an informal bargaining chip during negotiations with the interim CMJD, as mentioned in Chapter 4, the return programme was thus a formal pre-condition for the restoration of the political and financial support of the EU (Fresia, 2009: 50; Baila, 2016: 266).

returnee community in Rosso. The biometric transition is a determinant feature of the social dynamics into which the border regime intervenes, as it results in many being undocumented, despite having been born on Mauritanian soil.

The second grievance Babacar outlined is equally constitutive of the context in which the border regime intervenes in Mauritania, and thus of the form that this intervention assumes. This is the issue of land rights. He described how many who had been expelled in 1989 left behind land that was subsequently acquired by wealthy members of the *biḍân* class. As we saw in Chapter 3, this wave of expropriations was facilitated by a 1983 reform of the land tenure regime aimed at increasing the productivity of the Senegal River Valley (Crousse and Hesseling, 1994). The law abolished the system of customary collective ownership that had hitherto prevailed in the region and replaced it with a system of individual private property rights. This liberal abstraction was intended to make it possible for Mauritanian nationals without distinction to become property owners. But in effect, it served to expedite a restructuring of land ownership and social relations in the wake of 'the events'. In the words of the OECD (2010), the 1983 'reform ignores customary law and legalises expropriations, predominantly suffered by Mauritanian Fulani' (p. 10). The expulsions of 1989 accelerated these expropriations and consolidated the overhaul of the land tenure system envisioned in the 1983 law, thereby slotting this particular instance within a broader trend of accumulation by dispossession in the region, in which the border regime is no less implicated (Harvey, 2005; Cross, 2013).

Little wonder then, that in Babacar's view, 'the issue of land ownership remains to be resolved'. By the time he and his compatriots arrived through the return programme, neither the state nor the private owners of large-scale plots of land had any interest in returning it to what were largely Fulani subsistence farmers. Following the August 2008 coup, the European Parliament issued a resolution stating that it welcomes the return programme, but 'calls for the restoration of the rights of the refugees who have returned to Mauritania, and demands the return of the property confiscated from them' (European Parliament, 2009: 3). But much like the Rosso bridge and Nouakchott-Rosso road projects, this demand has suffered from a lack of follow-through. Instead, compensatory measures have been implemented as part of the tripartite agreement, such as the donation to returnees of livestock, the quantity of which varied between one and three cows depending on the size of the family, and rice and cooking oil. In the view of Babacar and the returnee community, these measures have been wholly inadequate. Abandoned infrastructural projects scattered around the camp served as further testament to this sense of tepid compensation. As we strolled through the settlements, Moussa pointed out a broken rice mill machine and a disused irrigation plot. Their abandoned skeletons appeared a sort of cruel and taunting caricature of the different components of the rice industry that now fuels Rosso, which has been largely erected on the back of this expropriation.

Having grown against this backdrop of displacement and dispossession, the local riziculture economy is today a key mediating site of social relations in Rosso. Thiam, of the local branch of FONADH, told me that a large portion of their activities concern addressing disparities in ownership between plots of land (interview, 16 July 2018). Also emphasizing the role of the 1983 land reform in shaping the aftermath of 'the events', he described how 'certain agribusinessmen now hold two, three, or four hundred hectares and don't do anything to develop them, while next to them are certain collectives, such as the returnees, who between them only have twenty hectares. They had land before and now have returned to nothing.' This is a consequence of what Thiam described as an 'anarchic' distribution of land following 'the events'. The translation of individual private property rights into informal clientelist relations has thus compounded the sense of exclusion and abandonment of the returnee community in Rosso.

Being a constitutive component of social relations in Rosso, this riziculture economy is also intimately tied to the subjective experiences of those illegalized by the border regime. Indeed, it is a key feature around which the lives of Ali Bakar, Soro, and many other migrant workers revolve in Rosso. We now turn to resituating their experiences more directly in relation to the history of dispossession and displacement that has left Moussa and Babacar abandoned on the peripheries of the town. If, as argued in the previous chapter, the assumption of a Europe-bound directionality needs to be abandoned in order for subjectivity's constitutive and transformative potential to be grasped, this is a crucial step to take. As we will see, what appear to be distinct structural processes of border externalization and accumulation by dispossession become tied together from the standpoint of this collective subjectivity.

Structural connections and collective subjectivity

The experiences of Ali Bakar and Soro in Rosso are structurally wedded to the changes in landownership that occurred in the wake of 'the events', and thus to the condition of Moussa, Babacar, and the community of returnees discussed above. This is because both Ali Bakar and Soro worked in different components of the Rosso rice industry – Soro in the fields and Ali Bakar in a rice processing factory. For different reasons, they were each on a temporary hiatus from this work and were killing time at the vegetable stall outside the border when I met them. As we have seen, the imprint of illegality fades from the surface of subjective experience for many migrants in Rosso. This in itself is not surprising; as scholars have long pointed out, migrant illegality is not an objective state, but a produced condition (De Genova, 2002; Menjívar, 2006; Squire, 2010; Fassin, 2011; Andersson, 2014a). In Rosso, however, certain features of illegality are observable even in the absence of this produced condition. Three such features can be highlighted, each one associated with the Rosso rice industry, namely labour exploitation, violence, and the 'temporal uncertainties' (Griffiths, 2014) ingrained within the migratory act of waiting. As we will see, these features are cumulatively connected, with exploitation of the bearers of labour power acting as the bedrock upon which the latter two are laid.

Firstly, despite the absence of the lived experience of illegality in Rosso, the labour exploitation to which it often predisposes migrants (De Genova, 2004;

Anderson and Ruhs, 2010; Mezzadra, 2011c) remains stubbornly persistent. When I met him, Soro had just finished a stint in the rice field, where he had had the ancillary task of clearing the field of birds and debris as the combine harvester combed through it. The rice field becomes a temporary home during the harvest period, with tents pitched there for field workers. The field, he said, was 120 hectares and belonged to a 'white Moor', who reportedly owned several such fields – a typical example of one of the 'agribusinessmen' to which Thiam of FONADH refers above. Now that the harvest had been completed,[7] Soro had come back to Rosso and was staying with Ali Bakar while waiting for a call to return to start work for a new sowing season.

Despite the absence of the labour disciplinary mechanism of deportation in Rosso, wage theft and exploitative working conditions are common for migrants working in the rice fields. Isaaka, the Senegalese community representative who lauded the local authorities in their treatment of migrants in the town, spoke less positively of conditions in the rice fields: 'the people in the fields, the seasonal workers, always have problems with the owners of the fields. Always. Because generally they're not paid by month; they're paid after the harvest. Sometimes the guy, he keeps the money, he doesn't pay them' (interview, 23 July 2018). Thiam, of the local branch of FONADH, also made reference to such instances, describing how 'agribusinessmen' employ recently arrived migrants without a contract and sometimes refuse to pay them at the end of the work term (interview, 16 July 2018). Soro was less candid when discussing such issues. He simply affirmed that some employers 'pay well, and others not so well', 'well' in this case being 50,000 UM per month, which comes to around €131. In any case, the very fact of his being precariously employed in the rice field underscores the exploitation of migrant labour in Rosso, notwithstanding the absence of migrant deportability.

Similarly, Ali Bakar had been working in a rice processing factory prior to opening the vegetable stall at the border. Like Soro, Ali Bakar's job was ancillary to machinery and entailed observing packaged bags of rice as they passed on the conveyor belt to ensure that none were torn or in otherwise poor condition. This he used to do over five 12-hour night shifts per week, for which he earned 60,000 UM per month. According to Thiam, addressing issues of abuse suffered by migrant workers in the factories and fields constitutes a large bulk of the work of FONADH in Rosso. This they do by raising awareness in Rosso about 'rights and duties' of migrants in order to better insulate them from vulnerability to abuse. They also intervene to mediate in individual cases of wage theft and other such disputes. While migrant labour exploitation thus remains present on the worksite in Rosso, Isaaka also emphasized there is a significant scope for informal community mediation and even recourse to legal authority, thanks to the absence of the threat of raids and deportations in Rosso.

7. This is how Soro described the timing to me. The main harvesting season is from October to December, but off-season harvesting also takes place in June and July, which is when we met in Rosso. Off-season harvesting is possible with an adequate irrigation system, the likes of which can be provided by the Senegal River.

Ali Bakar was on a break from his job in the rice processing factory when I met him. The reason for this involuntary pause highlights another quality of illegality that can be seen in Rosso, namely the violence to which migrants in the EU's Saharan buffer-zone are routinely subject (Bialasiewicz, 2012; Andersson, 2014a; Casas-Cortes, Cobarrubias and Pickles, 2014; Brachet, 2016; Stierl, 2016). While Soro's trajectory prior to arriving in Rosso was markedly more violent than Ali Bakar's, the latter's labour experience in the town was more brutal than that of his companion in the fields. He told me how one evening, upon noticing that one of the machines had stopped working, he reached in to fix it, only for it to restart suddenly while his hand was still in it. He lost his thumb. Retelling this story, he showed me the awkward angle at which it now stood from the point in his hand where it had been surgically reattached. Such injuries would appear to be commonplace; Thiam spoke of a case of a young worker from Guinea Bissau whom he had recently supported after he lost four fingers in a machine in a rice processing factory (interview, 16 July 2018). He was quick to emphasize that this was 'one case among many', and then recounted another example of a body washing up in one of the irrigation canals by the fields. While the direct violence that is overtly functional to the border regime is absent from social relations in Rosso, other less purposive but equally brutal forms characterize the migratory experience here.

Alongside violence there exists a third, more mundane, feature of migrant illegality that can be observed in Rosso. This is akin to what Griffiths (2014) describes as the 'temporal uncertainties' faced by deportable migrants, which captures aspects of violence and instability they experience, and tend to interpret in temporal terms. Soro had recently completed a stint in the rice field when I met him. He told me he would be returning in a day or two: 'tomorrow or the day after tomorrow. I'm just waiting for the call.' In the meantime, he idled with Ali Bakar at the vegetable stall, observing the comings and goings of the border. Two days later, he told me he was expecting to leave at 2 pm that day, but I found him 'still waiting' at the vegetable stall at the end of the day, with no call yet received. By the time I came back from Saint Louis a week later, he had left. A similar 'temporal uncertainty' is built into the experience of crossing the Rosso border, where the informal networks of patronage and access described above necessarily imbue the experience of waiting there with uncertainty. Irrespective of whether or not it is associated with the vulnerability of being undocumented, then, waiting with uncertainty is an integral and multifaceted component of the migratory experience (Conlon, 2011), one which appears as a constitutive feature of this experience in Rosso.

Despite the absence of deportability in the locale of Rosso, then, there remains a notable presence of certain experiences and phenomena associated with it. There are, however, important differences in some of their manifestations when shorn of the social condition of illegality. For example, a clear function and subtle intent can be detected in the violence and temporal slowing deployed by the border regime against illegalized migrants elsewhere in the EU's Saharan buffer-zone. In Rosso, however, these features appear less purposive and more incidental.

The interminable waiting experienced by those making the border crossing does not appear as an explicit tactic or function of that border, as is the case with the Ceuta and Melilla fences (Andersson, 2014b). Equally, Soro being made to wait for a phone call from his boss isn't necessarily a deliberate ploy on the part of the employer geared towards enhancing the productivity of his labour. The same can be said of the violence experienced by Ali Bakar and other migrant workers in the factories and the fields.[8]

These latter two features of illegality – violence and waiting – thus assume an incidental quality for migrants like Ali Bakar and Soro in Rosso. If we were to discard entirely with the historical context in which border externalization was placed in Chapters 3 and 4, we might be tempted to explain the presence of these features in terms of residues of illegality, whose overt function to the border regime dissipates with geographical distance from the EU borderline. This is a reasonable interpretation, but it also remains clouded by a two-dimensional focus on the border regime and its object of intervention. For this reason, it would inhibit the return of the gaze to which this chapter aspires. Situated against the social relations and structures in Rosso and the historical trajectory narrated in the third and fourth chapters, the undergirding factor of labour exploitation outlined above appears structurally linked to the mass expulsions of 'the events', and in particular to the dispossession and expropriation that occurred in their wake.

While the returnees remain discarded on the outskirts of Rosso, inadequately compensated for land that they lost, migrants like Ali Bakar and Soro work under dangerous and precarious conditions within the modern agricultural sector that has been erected upon this expropriation. From this vantage point, seemingly distinct processes of accumulation by dispossession and border externalization are tied together by what Mezzadra (2011a: 162) describes as 'a deep heterogeneity of subjective positions and experiences within the composition of contemporary living labor'. In this case, these subjective positions and experiences encompass violent pushbacks in the Sahara and interceptions in the Mediterranean, as well as the experience of being deported from one's country of origin and subsequently dispossessed of land there. The final section seeks to glean from this heterogeneous yet collective subjectivity an alternative perspective on the Rosso border to its reductive two-dimensional depiction by the EU border regime.

Returning the gaze

As we have seen, the dual quality of the Rosso border means that it cannot be reduced to its function as a constitutive dividing line of the interstate system, nor to the cross-border familial connections that traverse and exceed this

8. If anything, such workplace injuries act against the interests of employers, albeit in the infinitely more superficial sense of temporarily depriving them of surplus value, as distinct from grievous bodily harm.

territorial division. That these two contradictory qualities exist simultaneously is not surprising, and indeed is a reflection of the well-established dual capacity of borders to connect and separate (Walker, 1999: 447; Mezzadra and Neilson, 2013: 35). But this capacity takes on a particularly acute form in the case of a colonially inherited riparian border such as that which runs through Rosso, since communities have traditionally emerged and grown around the fertile banks of the Senegal River. This dual quality is alluded to by Marion Fresia (2009: 18) in her 'critical anthropology' of the asylum and humanitarian system set up on the Senegal River Valley in response to 'the events': '[C]onsidered as refugees by international law and therefore as foreigners, the Fulani of Mauritania nonetheless found themselves, from the point of view of the long history of the Senegal River Valley, on the land of their ancestors. Their close relatives were in fact counted amongst the Senegalese residing within their zone of reception.'

Within the context of our discussion, the EU-funded border infrastructure installed at the Rosso border can be viewed as a stand-in for what Fresia here highlights as the figure of the refugee as it is produced by international law. This is so in the broad sense that each of these manifestations of the international scale acts upon the socio-spatial environment of the Senegal River Valley, in a fashion that integrates it into a global Westphalian grid of national territorial ownership and belonging. On the other side of this coin lies the collective subjectivity of the historical residents of southern Mauritania and northern Senegal, whose spatial separation was initiated under the colonial shift in territorial organization, and is now sustained by these various instruments of the interstate system. While this border remains irreconcilable from their point of view, it also serves a productive purpose for many others within the underdeveloped and peripheral context of Rosso, as we will see.

The proximity of the past in Rosso

From the perspective of certain agents of the interstate system, and of the externalization process in particular, this territorial separation appears incomplete and woefully deficient. Adam, a former European delegation official based in Nouakchott, lamented the *laissez-passer* system in Rosso as being unfit for purpose. He explained that during the EU-funded border infrastructure project, technical experts envisioned a biometric system for town residents (interview, 5 March 2018). Under this plan, Rosso residents who make the crossing daily would have been registered biometrically with local authorities and granted access on the basis of biometric fingerprinting. But after the deterioration in relations between with the Mauritanian state in this domain, detailed in Chapter 4, this plan was shelved. In its stead, the analogue *laissez-passer* system remains in place. The shortcomings from Adam's point of view are manifold; people can pay each other for ID cards and then go through with the *laissez-passer*, despite looking nothing like the other person. The *laissez-passer* thus appears from this point of view nearly as a threat to the integrity of the international border that runs through Rosso, the validity of which is of course beyond question.

From a subjective position shaped by 'the events', on the other hand, the riparian border that runs through Rosso appears, at best, to be an anomaly. This was expressed clearly to me by Sy, who grew up in Kaedi, a border town on the middle Senegal River Valley (interview, Nouakchott, 7 March 2018). He was living in Nouakchott when his family was targeted during 'the events': 'I went to school, in 1989, and upon returning I found my mother and sister had been deported along with Senegalese!' This happened, he says, 'despite the fact that they had Mauritanian papers'. The memory of these experiences has led him to question the spatial effects of the border in light of the familial histories through which it cuts. He recalled to me how the border was navigated during his childhood in Kaedi: 'the Senegalese came on foot and the Mauritanians went. We swam! We swam to cross the river.' He then continued to reflect upon the incongruences between the colonial configuration of the river and the historical role it had played for those residing on each of its banks: 'these links are old. You know, it was colonisation that made it that way – you know, a river shouldn't be a border! That's why all along the river you see the same families. The brother on one side, the other brother on the other side. Same father, same mother!' From the perspective of the racialized margins of Mauritanian national belonging, then, the social contradictions of the colonial configuration of the Senegal River are manifestly apparent. This is all the more so when this subjectivity is formed against the backdrop of the racialized expulsions that were rendered thinkable by this configuration.

The social contradictions of the border are similarly apparent for some positioned on the Senegalese side of this international dividing line. I decided to wrap up my conversation with Isaaka by asking him to what extent things had changed in Rosso since 'the events'. His response is worth quoting in full:

> You know, for me, we – the population of Mauritania and Senegal – we have overtaken the action of the governments. Because, you know, the relations that you see there between the two banks, those are relations that date back thousands of years. So, while they talk about cooperation, blah blah! Us, we've already been cooperating. Integration has already been achieved for hundreds of years. You see an old man – he has one wife here, and his second wife in Rosso-Senegal. He spends two days here, two days over there. He has kids here – they're called Mauritanians – and kids over there – they're called Senegalese. Now, relation between Mauritania and Senegal at the level of government is one of distrust; each one distrusts the other. Well, maybe there are also people who really want peace to be a reality … But a Senegalese does not feel like a foreigner in Mauritania, and a Mauritanian does not feel like a foreigner in Senegal. That, it's done. It's already done … The only hindrance is the border.
>
> (interview, 23 July 2018)

It was suggested in Chapter 3 that the 'Senegalo-Mauritanian space' discussed by David Robinson (2000) (cf. Ahmed Salem, 2004; Ould Cheikh, 2004) was buried beneath a French colonial imaginary at the start of the twentieth century. But for

Isaaka it remains a simple lived fact, whose patency he subsequently expressed in a pithy refusal of the very category of migrant: 'you can't be a migrant at home.' From this point of view, the border appears a counterintuitive absurdity.

Isaaka further emphasized this absurdity by going on to juxtapose the social integration of the Senegalo-Mauritanian space that he had just illustrated with the long list of formal and informal obstacles posed for those crossing the border. Amongst these he listed the €50 worth of foreign exchange currency required of Mauritanians and Senegalese nationals entering one another's national territory, the vaccination certificate, the 500 CFA for the *laissez-passer*, and the informal duties and charges that are sure to be extracted at each point along the way. This maze of exploitative *formalités* thus compounds the sense of incongruence associated with the border for Isaaka. He then repeated: 'our only problem, the only hindrance, is the border! It shouldn't be this way. Really, it shouldn't be this way. Me, personally, my family is a triangle. One part of my family is Senegalese, the other part is Mauritanian, and the other part is Malian … so what are we? Well we're African!'

There is a clear resonance between Isaaka's sentiments here – 'it shouldn't be this way' – and those of Sy expressed above – 'a river shouldn't be a border'. Both also evoke personal familial ties and the Senegal River Valley's long history of interaction and exchange in their denunciation of the international border that now runs through Rosso. While Isaaka was born on the Senegalese side of this border and Sy on the Mauritanian side, the words of both express a deeply held sense of something being wrong with the very fact of this line of separation. At the same time, however, the wistful tone in which each of these sentiments is expressed indicates that this line of separation is today a spatial fact. Moreover, it is one from which many others in Rosso derive material benefit.

The productivity of the border and the subjectivity of living labour

Being situated just outside the border, Ali Bakar's vegetable stall was able to avail of a reliably steady stream of two-way traffic coming to and from the crossing; all of the customers who approached his stall while I was there were cross-border travellers. This speaks to the productive nature of the contemporary border; the dense and opaque informal economy outlined above make it a veritable means of gaining a living for those sufficiently shrewd and socially connected to work within it. While for Sy and Isaaka the border is a corrupt, extractive labyrinth constructed atop a historical aberration, the informal economy at the border is also a reliable livelihood source for many in this peripheral locale.

As we have seen, the land expropriations that took place following 'the events' indicate a structural link between the racialized violence of the border and the migrant labour exploitation in the Rosso rice industry. From the point of view of Ali Bakar and Soro, however, this labour exploitation is not experienced in overtly negative terms. Indeed, Ali Bakar was eager to return to the factory, as he would there earn a regular monthly salary of 60,000 UM rather than the piecemeal income that he received at the vegetable stall: 'business isn't good. Me, I don't have a wife and kids here. So it's better that I work in the factory where I get a monthly salary. Half I can send to the family back home and half I keep for expenses here.'

The security of a regular income is particularly appealing compared with the dearth of opportunities in his home region engendered by multiple intersecting crises: 'before, there was rice there, like the rice fields here. But now there is very little rain. And since the war in the north with the jihadists, everything is ruined.' Soro also spoke of the violence that had engulfed his home region of Mopti, disparaging the jihadists as 'bandits'. For him, too, the rice fields in Rosso were a more attractive option than Nouakchott, where he had spent months looking for work: 'it's hard in Nouakchott', he said to me as we sat at the stall, 'it's better here. There's more money.' By many standards, the conditions of the different components of the rice industry in which Ali Bakar and Soro work are low-paid, precarious, and dangerous. But from the vantage point of subjectivities forged through the violence of the EU border regime, jihadist insurgency, and climate breakdown, such exploitative wage labour appears a comparatively appealing source of stability and security.[9]

On the one hand, then, the border is an externally imposed historical aberration, a category error whose social contradictions are manifestly apparent for those who were displaced and dispossessed during 'the events'. From this point of view, the gruelling experience of negotiating with police and *tchib-tchaba* intermediaries as one crosses the border is but further evidence of its incongruence with the environment in which it operates. This incongruence is reinforced by the physical and technological infrastructure now installed at the border by the externalization process. On the other hand, for those positioned to benefit from this informal economy at the Rosso border, there is no such incongruence, as the presence of international migration management procedures is a site from which material benefits can be reaped. And while the restitution of land remains a key demand of those who were deported during 'the events', for migrants like Soro and Ali Bakar whose prior experiences encompass border violence, ecological crisis, and civil war, the riziculture economy that has grown from this dispossession provides a meagre level of stability, one that is likely enhanced by the absence of deportability for migrants living in Rosso.

Clearly, the collective subjectivity of living labour in this context is wrought with internal contradictions. Nonetheless, it is shaped by a colonial legacy of racialized territorial belonging and a related contemporary process of accumulation by dispossession. The former can be seen in the mass expulsions of Mauritanian citizens during 'the events', and the latter in the land expropriations that followed in their wake. Border externalization necessarily intervenes within the contradictory set of social relations and structures endowed by this particular history, and is, in turn, shaped by them.

9. Violent conflict has been argued to function as a process of primitive accumulation through its generation of displaced communities who become vulnerable and thus exploitable labour forces (Cramer, 2008).

Conclusion

From the two-dimensional vantage point of the EU border regime, Rosso appears either as a node on an invariably Europe-bound northward journey or as a point through which deportations aimed at quashing such journeys can be enacted. This chapter has nuanced this homogenizing depiction by ethnographically situating the externalization process in relation to pre-existing social relations, structures, and subjectivities in this particular locale. It started out at the Rosso border crossing, where the material and technological infrastructure furnished by the EU inserts this site within a regional deportation network and a global regime of migration and border management. At the same time, however, Rosso's peripheral condition in Mauritania creates a zone of informality at the border, which endows the migration management protocol enacted here with an entirely distinct surplus extraction function. We also saw that the *laissez-passer* system to which Rosso residents are subject acts as a legal compromise between the international border at Rosso and the familial and economic dynamics of the social life that traverses it. While formally distinct from border externalization, in practice these facets of socio-spatial life in Rosso act upon it and come to constitute it in its unfolding.

Turning more directly to social relations in the town, we saw that illegality fades as a subjective experience for Ali Bakar, Rosso, and other migrant workers in the town of Rosso. In order to discern what comes in its stead, we moved farther from the border still, to the outskirts of Rosso, where those who were deported during 'the events' are now discarded, dispossessed of their former lands. This issue of land restitution drew attention to the violent displacement and dispossession underpinning the rice industry in which Ali Bakar, Soro, and countless other migrant workers in Rosso are employed. From this perspective, the conditions of Ali Bakar and Soro in Rosso appear structurally wedded to the racialized violence and dispossession of 'the events'. We then came back to look once more at the Rosso border, though this time from the vantage point of the collective subjectivity of the various actors encountered during this ethnographic exploration.

This chapter has therefore retained Sandro Mezzadra's (2011c: 137) call to analyse the 'transformations of contemporary capitalism from the point of view of living labour and its subjectivity'. But in doing so, it has moved away from an exclusive focus upon 'irregularity [as] an ambiguous condition that forms a key political stake in contemporary social struggles around capital and migration' (Mezzadra, 2011c: 122). Indeed, it has unsettled the category of irregularity/illegality altogether, suggesting it to be not just a produced condition in itself, but one which may be a mere surface-level reflection of more deep-seated processes. This is an important observation, but its immediate relevance to the externalization process may not be apparent from the particular perspective of the locale of Rosso. Returning the gaze has shown externalization to be bound up in regional histories of displacement and dispossession, but it is less clear whether and how the agency of the collective subject position that emerges from these entangled histories constitutes and transforms the externalization process, in the manner of the socio-spatial dynamics of the Rosso border. In order to discern the relationship between the more general condition underpinning illegality and EU border externalization, something more than returning the gaze on the border regime is required. This task will be undertaken in the following chapter, which moves inwards from the borders to the capital city of Nouakchott.

Chapter 7

BORDERING POSTCOLONIAL CAPITALISM

In the same Nouakchott public square in which we encountered Pierre at the start of this book, people wait out the days in search of sparse and precarious employment. The high, barbed-wire-adorned walls of the French embassy loom just beyond this roundabout, a totem of the European border walls and fences that shape the experience of Pierre and so many others here. The north end of the square is overlooked by a large expensive apartment complex, home mostly to international military and diplomatic staff, and a newly constructed hotel. Numerous people here recounted to me how the owner of this hotel enlisted the police and the municipality to clear them out of the square while it was under construction in 2016. Several were deported to Rosso-Senegal during these raids. The building was then left empty and unused for over a year, seemingly forgotten by its owner. But the expulsions that accompanied its construction are still fresh in the minds of everyone here, as they acted as a turning point in their relationship to this place. Before the authority of the state was deployed against them, they were free to carry out their car-washing business in the square itself. Ever since, they have been relegated to a laneway adjacent to it. On the south side of the square is the walled-off perimeter of Nouakchott's only Christian church. To those in the informal economy in operation here, the square and its surroundings are simply known as the Church.

In the previous chapter, we examined elements of the EU border regime's external operations from the perspective of its extreme periphery, the Senegal River Valley town of Rosso. From this vantage point, the 'irregular migrants' whom the externalization process treats as isolated objects of intervention appeared equally enmeshed within the region's history of displacement and dispossession. A similar entanglement between migration control and the violent exclusions of contemporary capitalist development is evident in the context of the 2016 deportations at the Church, which was, after all, one of an urban clear-out operation instigated at the behest of a property owner. What is the relationship

Portions of this chapter appear in an issue of *Movements: Journal for Critical Migration and Border Studies* (Ould Moctar, 2018).

Figure 4 Workers gathered in front of a wall at the church in Nouakchott. Photo by author.

between the history of accumulation by dispossession charted in the previous chapter, the urban-clear out that those at the Church experienced in 2016, and the deportations that subsequently befell many amongst them? Put in more general terms, how do rural dispossession processes, the exclusions of the urban informal economy, and the illegalization of migration interact with one another? This chapter charts an intertwinement of illegalization, racialization, and economic abandonment in the urban informal economy of Nouakchott. In doing so, it shows how the historical processes detailed in Chapters 3 and 6 converge with the contemporary exclusions of the capitalist economy in Mauritania to create a social condition from which the border regime can draw in its illegalization of those suspected of being on the move to Europe.

In theorizing this condition, the chapter draws from a blend of critical scholarly traditions which have in distinct ways theorized the racial and colonial underpinnings of capitalist development in the Global South (Sanyal, 2007; Aung, 2018; Bhattacharyya, 2018; Mezzadri, 2022; Bhattacharya, Kesar and Mehra, 2023). While diverse in regional focus, this scholarship has in common a problematization of the figure of the 'free' wage labourer that has informed more orthodox Marxist definitions of capitalism, and, relatedly, an interrogation of the problem of absorption of those who are dispossessed, cast out of the development process, and left without any readily available means of survival. Unlike in Europe, where those rendered surplus relative to capital have historically emigrated en masse thanks to a *de facto* free movement regime in place for many citizens of the colonial metropole, those in a similar position today are faced either with a violent regime of restrictions to movement, or abandonment and confinement in the slums and informal economies of the Global South (Davis, 2006).

Those at the Church in Nouakchott are but one small segment of this global class who have been 'expelled from the agricultural sector and at the same time excluded and segregated from the glittering, modern urban economy' (Sanyal, 2007: 46). If the previous chapter offered insight into such processes of expulsion on the Senegal River Valley, this one brings to light these dynamics of segregation from the modern urban economy. In detailing the experiences of illegalization and border violence of many in this site, the chapter shows how the border regime intervenes within the process of producing these surplus populations. As we will see, there are some uncanny parallels between the spatial mechanisms of controlling the movements of surplus populations at the level of the urban informal economy and the externalization measures to which many within this self-same group have been subject.

The chapter develops this point in four stages. Firstly, we see that the illegality produced through externalization at a regional scale manifests in Nouakchott. Discussion of further illegalization at the hands of Mauritanian security forces, however, soon draws attention to its contextual contingency, which in turn gives rise to an inquiry into the forces that determine who becomes illegalized and why. In principle, any policy formula or analytical investigation concerned with migrant illegality is premised upon a national/other separation. The rest of the chapter goes on to demonstrate two ways in which this separation comes apart in Nouakchott. The second section does so by examining the ongoing ramifications of the colonial racialization of territorial belonging, which means that the boundary between inclusion and exclusion does not always correspond to that between national and non-national in the informal economy. The third section moves on to analyse this informal economy more directly, demonstrating how all who move through it find themselves in the position of being surplus relative to the needs of capital, irrespective of nationality. In this light, externalization appears to be a particularly violent and visible aspect of a deeper process of bordering, containing, and dispersing those in this structural position.

Unlike the previous chapter, then, which sought to draw out the contemporary reverberations of a particular localized history, this one strives towards a more global frame of reference, despite being focused on the particular site of Nouakchott. In other words, this chapter moves beyond Chakrabarty's (2001) call to 'return the gaze' which guided the analysis in Rosso. In adopting a global frame of reference, it is oriented by the non-Eurocentric universalism advocated by Samir Amin. Through a concluding exploration of how those on the urban margins interpret their everyday reality and experience, the chapter finds some speculative ingredients of such a universalism.

On the production of illegality in Nouakchott

'We agreed to keep the immigrants here!,' the coordinator of the national migration strategy exclaimed during a diatribe against the EU's ingratitude for this favour on the part of the Mauritanian state (interview, Nouakchott, 29 March 2018). There is an accuracy to this depiction of Mauritania as a sort of holding zone for migrants unwanted in Europe. As sea departures diminished following the externalization measures deployed from Nouadhibou post-2006, detention centres, police cells, and urban pockets of deprivation in Mauritania swelled. It was during this period that, as Philippe Frowd (2018: 164) observes, the security focus shifted from 'who is crossing the border and how do we stop them?' to 'who is here and how do we track them'? As we have seen, from 2012 onwards, this security focus was embodied in a new biometric residence permit for non-nationals. As capital city and home to a significant portion of Mauritania's non-national residing population, Nouakchott soon became a prime site of enforcement of the residence permit. As detailed in Chapter 4, national police and gendarmerie would periodically descend upon migrant neighbourhoods and community spaces, with migrants who could not flee fast enough being rounded up and detained, and many amongst them being subsequently deported. These experiences of illegality and deportability in Nouakchott are, however, also expressions of a regional phenomenon of border violence associated with externalization, which are embodied and reproduced in the Mauritanian capital city. In what follows, we explore these regional dimensions of border violence as well as its sustenance by security forces in the city itself.

Regional border violence embodied

Follow a road leading north from the Church, and one soon comes across another informal car-washing site. Flanked on one side by the French embassy and on the other by the G5 Sahel Permanent Secretariat, the road eventually gives way on the left to a luxury hotel. Just beyond its compound wall gate, a group of young Malian and Mauritanian men have commandeered a disused strip of land. A stretch of sandy terrain spanning 20 yards or so between the road and a crumbling wall, this space provides an ideal area from which cars can be hailed and car washing carried out. Behind the wall, such everyday essentials as tea-making sets, car-washing materials, and prayer mats can be stored.

It was here that I met Zein, a Fula youth from northern Mali. Energetic and self-assured, he was set on reaching Europe, which he repeatedly described to me as 'his dream'. On slow days, debates about everything from political unrest in the Sahel to social inequalities of Mauritania would arise. Of course, the prospects and methods of circumventing the vast border infrastructure designed to keep people like Zein outside EU territory often also figured in such discussions. Zein listened entranced on one such occasion, as a companion regaled stories of the clashes at Ceuta and Melilla to a captive audience. 'I'm going there. 2018, *inshā' Allah*', he said, taking a pensive drag on his cigarette and gazing determinedly into the distance. To my surprise, however, a few weeks after this conversation had taken place it was not Zein but another quieter member of the same group who had left for Morocco to get a boat to Spain from Tangier. By the end of my fieldwork period, Europe had for Zein not moved beyond its mystical dream status.

For others at such sites, the prospect of reaching Europe was equally distant, albeit as a past memory. Boubacar, for instance, was also an ethnic Fula, but was, in every other respect, Zein's polar opposite. Quiet and reserved, he was not prone to protracted monologues or offering unsolicited opinions, but he did welcome my company, always offering a seat on his curb to share parts of his life story. He was older than many others I met on Nouakchott's urban margins. He had attempted to reach Europe in 1998 on boat that left Libya for Malta, only to be intercepted by the Tunisian Coast Guard. Everyone in his group had paid $3,000 for the boat, but he found himself deported back to Senegal. 'We just wasted our money,' he said to me with a remorseful clarity of hindsight. Boubacar felt he was now 'too old' for the journey to Europe and was resigned but resolved to earn his keep in West Africa.

More recent encounters with the EU border regime in North Africa are also dotted throughout this landscape in Nouakchott. Souma, whom I met at the Church, had been deported to Guinea from Libya after leaving its shores on a boat that was intercepted by the Libyan Coast Guard in April 2018. He came to Mauritania shortly afterwards. 'The Libyans are nasty,' was all he would say of this experience. Another young man from Mali told of his time in Melilla trying to cross the Morocco–Spain border fence. He was removed from Melilla on upwards of forty separate occasions, in what Lorena Gazzotti and Marie Hagan (2020) describe as internal 'arrest and dispersal' operations that Moroccan authorities wage against migrants in Moroccan cities upon Spanish request. And as we saw in the introduction, Pierre eventually arrived at the Church after being deported from Morocco to Nouadhibou, following a thwarted attempt to cross into Spain via the Melilla border fence, while his car-washing companion Khalifa had been abandoned in Algeria's southern borderlands.

While sharing an experience of EU border violence, Pierre and Khalifa bonded in Nouakchott over the informal car-washing profession. As scholars of migration and borders often point out, border regimes are, in addition to being violent an exclusionary, often productive of new relations and spaces (Anderson, Sharma and Cynthia, 2009; Novak, 2022). In Nouakchott, the border violence experienced by many at a regional scale forges new livelihood strategies in the urban sites to which people are consigned. Khalifa explained how he and Pierre came to work together: 'Since he's new and he doesn't know the job too well, I show him the ropes and we share the money.' There was a hierarchy to this arrangement, however; being a newcomer, Pierre would often be tasked with standing at the roundabout trying to attract new customers. Khalifa, having already built a customer base, had the relative luxury of waiting for clients to come to him, and of sharing in the spoils from those Pierre succeeded in attracting.

Similar such relationships are replicated across Nouakchott's informal car-washing and labour-pick up points, giving rise to further relations and practices. At the car-washing site in which Zein worked, Fatima, a woman of Guinean and Senegalese parentage, came on a daily basis with everything required to prepare and serve lunch to upwards of thirty people. Typically consisting of a plate of the

Senegalese rice dish *Thiéboudienne*, her lunches were eagerly enjoyed by all here, not least Zein, despite his sheepish pleas that he would pay the 400 UM that Fatima charged at the end of the week. She would typically react with a knowing roll of the eyes, before handing him a loaded plate.

If border violence and exclusion are productive of new relations and practices, they in this case produce what Gargi Bhattacharyya (2018: 55) calls 'the emergency survival tactics of sustaining human lives alongside uncertain or non-existent access to waged work.' As she suggests, these tactics often only ensure a bare minimum of survival, and for this reason cannot be celebrated. But they also demonstrate that the reproduction of social life cannot be reduced in its entirety to the logic of capitalist production. If it could, Fatima would have no reason to shrug off Zein's coy deferrals of lunch payment.

Within multiple such pockets of urban deprivation in Nouakchott, then, there is a dovetailing between the experiences of border violence detailed above and informal spaces defined by limited access to wage work and precarious tactics of collective survival. As we will later see, this overlap is no coincidence. Firstly, however, it must be noted that these experiences of illegality emanating from elsewhere in the region are also reproduced in Nouakchott.

The ebbs and flows of deportability

One evening in 2015, Barry was sitting with friends drinking tea in the *5ème* district of Nouakchott when the police van pulled up. There was no discussion, he says, no questions about papers. They were simply bundled up and taken to the police station in the Nouakchott neighbourhood of 'Baghdad', which had become an effective migrant detention centre (cf. United Nations Human Rights Council, 2016). Here he spent two days, before being brought to Rosso. Legally, Malian nationals are to be deported to their home country via the border crossing at Gogui, in the east of Mauritania. Nonetheless, Barry found himself in Senegal after having been deported through Rosso. Like in other such instances, the police who detained Barry barked the phrase '*séjour*!' as the raid unfolded. As we saw in Chapter 4, the introduction in 2012 of the biometric residence permit – the new *carte de séjour* – perpetuated the deportability that had originally been introduced six years earlier, ingraining it across various migrant community spaces in Nouakchott.

The link between this new social reality and the externalization agenda was widely emphasized by local media and human rights organizations at the time. 'With the controls imposed by the European Union', Mamadou Thiam (2016) observed in an investigate report on the policing of the residence permit in Nouakchott, 'repression of illegal immigration has intensified.' As Barry's experience suggests, the approach in Nouakchott appears to have been to round people up first and ask for documents later. Back in 2012, in the words of a Malian community representative (interview, Nouakchott, 13 December 2017), 'as soon as they saw a group, they rounded everyone up! They don't ask you anything, they round you up and take you to the police station.' Those who happened to have a residence permit would spend a few hours or the night in the station, before being let go. The rest, according to this representative, were likely to be deported.

For some who meet this fate, however, the possibility of returning to Mauritania was not ruled out. Barry, for instance, worked for two months in Dakar before making his way back to Mauritania in early 2016. Like Malata in Nouadhibou, he re-entered Mauritania through Rosso, legally authorized to spend another three months on state territory. As we saw in the previous chapter, the informal structure of bribes and favours in operation at Rosso inhibits the efficacy of the externalization infrastructure at this crossing. A similar such structure also often precludes deportation in Nouakchott. According to Barry, even when you are picked up without a residence permit, 'they will leave you alone' if you pay 1,000 UM. Another individual told of being twice detained in Nouakchott for lacking a residence permit, and evading deportation on each occasion thanks to the intervention of his employer.

Yousef, also from Mali, was arrested after being asked for the permit. He was detained and threatened with deportation unless he paid a sum of 10,000 UM. He rang his sister's son, who lives in Nouakchott and speaks *ḥassâniyya*, to come bail him out. After some negotiation, he was released late in the night, 'without enough money to get home or even buy cigarettes'. Myriad other such accounts emerged over the course of fieldwork, which collectively have implications for the efficacy of externalization's production of illegality. Once forged as a policy category and embodied as a subjective experience (Andersson, 2014a: 124), one can seldom negotiate oneself *out* of illegality, but negotiations can and do take place around how it is lived, as the above experiences suggest.

Whether or not such negotiations even have to occur in the first place is, furthermore, conditioned by broader political dynamics. During the time I spent in Nouakchott, and in stark contrast to Nouadhibou, people's responses across multiple sites in the informal economy whenever I asked about the *raffles* and *ramasses* surrounding the residence permit were often that the *séjour* affair has 'died down'.[1] There are a number of factors underpinning what has transpired to be a momentary lull in migrant policing in the city between 2017 and 2019. In May 2016, a Malian national died fleeing the construction site in which he was working after security forces arrived checking for residence permits, provoking an outcry from human rights organizations and a rebuke from Malian consular authorities (Wane, 2016). At the same time, an inquiry instigated by the new head of the General Directorate of National Security (DGSN) into the extent of bribery and other forms of malpractice by police officers was reaching its conclusion (Sileye, 2016). On the basis of the findings, the DGSN head in November ordered a temporary suspension to migrant policing operations in the city. The AMDH seized this opportunity the following month to reiterate their calls for a halt to raids and arbitrary arrests of 'Sub-Saharan Africans'. They would later credit civil society lobbying for the relaxation of the deportation campaign during this time (AMDH, 2019).

1. '*Diminué*' in French.

This juncture generated a brief sense of cautious optimism on the part of migrant community representatives and AMDH officials with whom I spent time in Nouakchott. The new DGSN head even agreed to attend a seminar in December 2017 about the problems faced by migrant communities in the city. At the event, he listened attentively to the demands for an alleviation of the conditions required to obtain the residence permit, he reiterated that he had ordered the *raffles* in migrant workplaces to be suspended and requested that each migrant community compose a list of issues affecting them for it to be considered by the security authorities. All of this led the CCMM convenor, at the next meeting of migrant community associations in Nouakchott in early 2018, to inform members that 'there is a will to resolve the problems of migrants in Mauritania', and that 'he [the DGSN director] wants the communities to be involved in the management of problems that they face'. Eight months later, however, this sense of good will appeared to have totally evaporated, as raids and deportations were once again picking up pace in the city (interview with CCMM convenor, Nouakchott, 15 August 2018).

While the prevalence of deportability across the urban landscape therefore ebbs and flows in accordance with broader political and institutional factors, the social life upon which the condition of illegality has recently been grafted of course remains present in the city. And for the vast majority, the material fact of lacking a residence permit, and of thus being 'illegal', remains unchanged. In other words, the condition of illegality and its derivative state of deportability are but elements of a more general state of social being. While border externalization depends upon its production of illegality to justify its existence, the social life and relations from which it draws this existential validation cannot be reduced to this artificial condition. Rather, it is imposed upon particular swathes of people who come to be illegalized due to their structural position within the societies into which the border regime extends.

So, if illegalization is but the tip of a larger iceberg, what does the rest of it look like? The remainder of this chapter dives beneath the surface to explore this collective structural position. It does so by highlighting two features of the social condition from which illegality is drawn. The first is the ongoing legacy of the colonial racialization of territorial belonging, whose localized reverberations on the Senegal River Valley were explored in the previous chapter. And the second is the urban informal economy of Nouakchott – a prominent site into which illegality is periodically infused – as well as the broader machinations of postcolonial capitalism that produce it. As we shall see, for distinct reasons, national lines of distinction become strained and collapse within each of these processes, bringing the social condition upon which the border regime imposes illegality into sharper focus.

Racialized territorial belonging and national distinction

The previous chapter has demonstrated that the EU border regime intervenes within older racialized social structures in Mauritania. But it is also itself part of a distinct historical structure of European white supremacy (Vives, 2011;

Lemberg-Pedersen, 2019; Stierl, 2019). The disproportionate degradation of Black lives that occurs as a result of EU cooperation with states south of the external borderline makes externalization a structurally racist endeavour (De Genova, 2017a: 22; Gazzotti, 2019: 426). In the Mauritanian context, the necessary racism of the EU border regime nourishes and reshapes existing racialized structures and discriminatory practices, as it has in Morocco and Libya (Andersson, 2014a: 125; Brachet, 2016: 282). Pierre gave voice to the weight of these layers of racialized hierarchy as he described the experience of being forcibly removed from Spain and then from Morocco. 'You know, we with black skin, there's no regard for us.' White supremacy, in his view, is ineradicable; 'it goes to the foundations'. Pierre's perspective is clearly a fatalistic one, and understandably so given his experience. But his point about foundations also refers to a convergence of distinct racialized hierarchies that occurs through the externalization process. As this convergence unfolds on the ground in Nouakchott, the postcolonial legacy of racialized territorial belonging disrupts the legal distinction between national and non-national which underpins the residence permit.

The racial blurring of national boundaries

At a regional trade union conference on the dynamics of migration held in Nouakchott in February 2018, in Nouakchott, discussion turned towards a recent incident reminiscent of that which had catalysed the racialized violence of 'the events' twenty years previously. Just a few weeks prior to the conference, the Mauritanian army had shot and killed a Senegalese fisherman who had strayed beyond designated fishing limits on the Senegal River (Maillard, 2018). The subsequent backlash against Mauritanian *biḍân*-owned businesses in Saint Louis invariably stirred memories of the all too similar, though much more widespread, targeted pogroms in Saint Louis and Dakar that unfolded during 'the events'. This time round, however, the leadership of each country was inclined to urge calm rather than stoke tensions,[2] and the violence didn't escalate further. But it did impel reflections on the history of 'the events' and their associated legacies of discrimination and exclusion. It was in this context that a state official at the conference professed to being proud of the biometric residence permit system, because 'now, no one is targeted because they are such and such. It's because they haven't respected the law.'

There is a conspicuously neat schematic break implied within this statement, which suggests that in the era preceding the residence permit, people were 'targeted' for reasons that were either arbitrary or discriminatory, as opposed to the national logic which nominally informs the policing of the residence permit. In the view of a Malian community representative, however, while the

2. An impulse likely driven in no small part by an agreement between the two countries on a highly lucrative liquified natural gas project located on an offshore gas field straddling the Senegal-Mauritania maritime border.

era preceding the residence permit was indeed characterized by arbitrary abuses, today 'the cases of abuse continue, and the residence permit has been added to all of that'. Moreover, it would appear that a parallel logic continued to inform how permits are issued at the time of my fieldwork. 'If you go to the civil registry', a member of a community association told me during one of the AMDH meetings, 'it's exclusively Europeans, Chinese, people who work in big companies. Not Sub-Saharans.' The necessary corollary is that it is exclusively 'Sub-Saharans' who are targeted during raids and deportations. As one Senegalese journalist observed to me (interview, Nouakchott, 26 February 2018) while discussing the findings of a parliamentary inquiry into the raid and deportation campaigns in 2012: 'it's the black Africans who were being deported. The Moroccan nationals, for example, we've never seen them being chased left and right.' If racism is inherent to the discourse of the securitization of migration (Ibrahim, 2005), it would appear all the more so in its practice.

For this reason, the policing of 'illegal' migrants in Nouakchott necessarily spills over into racist targeting of Afro-Mauritanians. Numerous accounts emerged over the course of fieldwork of police threatening Afro-Mauritanians with arrest and deportation. Iksander, a young man born in Nouakchott to Malian parents for example, responded thusly when I asked if the *ramasses* were ongoing in his neighbourhood:

> Yeah, a bit at night. They make you pay two- or three-thousand UM if you're a foreigner. If you refuse, they deport you. They even stopped me the other day. But me, I'm not going to pay. I started shouting in *ḥassâniyya*; 'this is illegal', and all that. Finally they said 'you talk too much, get out of here.' But lots of other people are scared. They just pay.

Having been born in Nouakchott and learned the *ḥassânîya* dialect of Arabic, Iksander is perhaps a bit more defiant than the average migrant worker in the city. But despite having been born in Mauritania, he has never been able to acquire civil registry documents. When I asked why this was the case, he responded in a matter of fact tone: 'here they don't give papers to blacks. You go in and they ask you for the papers of your great, great, great grandfather.' As seen in Chapter 4, the biometric overhaul of the civil registry through which the residence permit was introduced resulted in a wave of protest on the part of Afro-Mauritanian campaign groups. For individuals like Iksander, the biometric revamp has resulted in him being in the peculiar state of being undocumented in his home country, and therefore suspected by authorities of being 'illegal'.

There is thus a consistent overlap in Nouakchott between the policing of irregular migration mandated by border externalization, and a deeper postcolonial history of targeting and deporting racialized Mauritanian nationals suspected of being 'foreign'. Zein gave succinct, if passing, expression to this overlap during one of his characteristically polemic social analyses: 'the *séjour* affair is just racism; even Mauritanian blacks don't have papers!' Indeed, gathered alongside one another in the shade of the walls at the Church and elsewhere in the city are

Malians, Guineans, Senegalese, as well as Mauritanian Fula, Soninké, and *harâtîn*.[3] In the urban informal economy of Nouakchott, the legal line dividing deportable non-nationals from non-deportable national citizens is overlain with the colonial legacy of racialized territorial belonging, which, in turn, informs how things play out when migrant policing raids sweep through the city.

That said, the national/other divide is not merely an institutional fiction, lacking in any social or subjective substance. As we have seen, the racialization of territorial belonging blurs this divide in the eyes of security forces tasked with carrying out migration control operations in the city. But at the same time, those racialized through these policing operations often reproduce this same divide amongst themselves on a daily basis.

Reproducing national distinction

In the shade of a tree in the laneway on the north end of the Church, Yousef, the Malian whose experience of evading deportation was related above, sat with a Mauritanian Fula called Moussa. The day was hot and work was slow. As often happens at such moments, the conversation turned to the reasons underlying the social condition of those here. Moussa, for his part, couldn't understand how a country so rich in natural resources could yield such crushing poverty for its people; 'here there's nothing. There's iron, there's gold, there's fish; but there's nothing.' Yousef, meanwhile, criticized the inequalities that are born of this resource distribution, observing that we are all equal in the eyes of God. But opinions differed when Moussa reframed these inequalities along ethno-racial lines, speaking about how Black Mauritanians were oppressed by 'white Moors'. Yousef was sceptical in this regard, insisting instead that Afro-Mauritanians 'are not real Mauritanians'. His sense of divinely ordained equality notwithstanding, Yousef – a migrant worker from Mali – thus holds a sentiment reminiscent of that which accompanied the expulsions of 'the events', and which people like Iksander continue to encounter at civil registry centres today.

Such is the potency of national distinction within certain pockets of the Nouakchott informal economy, it even drives cleavages between members of the same ethnic group. Boubacar, whose experience of being intercepted by the Tunisian coastguard was described above, is an ethnic Fula from Senegal. Over the course of conversation, he expressed the following opinion of his Mauritanian Fula counterparts: 'They're worth nothing here. They're behind. And they always give foreigners a hard time. They say that foreigners have ruined the country. They say that this is their country, but it's not their country! They only collect rubbish and that type of thing. And they give us a hard time.' Boubacar's point about the structural marginalization of Mauritanian Fulani is a valid one, as Riccardo Ciavolella (2010) has shown in his 'anthropology of the margins' of the

3. The presence of formerly enslaved *ḥaraṭîn* in this site also speaks to the ongoing legacy of precolonial modes of racism and racialization in Mauritania.

Mauritanian state. For this reason, I suggested that the Fula of Mauritania give people like him a hard time because they too are having a hard time. 'Yes, they are having a hard time,' Boubacar agreed, before continuing to decry their attitudes to foreigners. 'But you speak the same language, don't you?' I asked. 'We're brothers!' he affirmed, 'but they give us a hard time.'[4]

That Boubacar should feel such antipathy towards fellow Fula who happen to sit on the opposite side of a national dividing line is, perhaps, not so surprising. The divide may have been imposed during the colonial era, but this does not render it any less 'real'. This comes across clearly in William F. S. Miles' (1994) study of the colonial fault-line between French and British West Africa that was imposed upon the Hausa people. 'As surely as the colonial boundary was superimposed upon the Hausaland,' he observes, 'so have two new identities' (Miles, 1994: 62). In the Mauritanian context, such sentiments are no less pronounced on the national side of this divide. Alex, a soft-spoken Afro-Mauritanian car washer, often prided himself on rising above disputes, on not becoming bogged down in the petty acrimonies that often shaped relations at the Church. Nonetheless, he held migrant workers there responsible for the low rates of pay that had become the norm over the years: 'all here are Malians, Guineans, Senegalese. Mauritanians are maybe only 20 per cent. Them, they work for nothing – they've ruined everything.'

Life had not exactly been kind to Alex. Having completed secondary and third-level education, he found it impossible to find work in Nouakchott, and was of the firm opinion that informal interventions, and not formal qualifications, are what count when it comes to getting a job. For this reason, he tried to reach Spain in 2014, but was arrested in Santa Cruz after having made it to Tenerife, and from there deported to Nouadhibou. Next time, he told me, he would save money to get a visa to migrate legally, like Camara, whom we met in Chapter 5. In the meantime, however, he is forced to compete with migrant workers for customers at the Church.

Youslem, a Mauritanian Fula whom I met at the Church, voiced similar sentiments after throwing himself on the ground in dejection after trying his luck in the crowd that had gathered around the car of an employer. 'The guy said he doesn't hire Mauritanians,' he told me, before going on to claim that he used to earn 60,000 and 70,000 UM per month as a domestic worker back when he first came to Nouakchott. Now, he was only earning 25,000 to 30,000 UM, a change that he attributed to migrants. They work for nothing, he said, and as a result 'they have ruined everything'. For Boubacar, the hostility of Mauritanian Fula like Alex and Youslem in his regard is the result of a superiority complex, borne of their own national marginalization. Youslem, however, views migrant workers such as Boubacar to be the cause of his structural marginalization. When slotted within the volatile pricing and hiring dynamics of the labour pick-up at the Church, national schisms thus acquire a hostile potency.

4. Originally '*Ils nous fatiguent*' in French.

While the line between nationals and non-nationals is blurred in the racializing view of the state, it is crystal clear from the perspective of those subject to this racialization, who are forced to share the meagre spoils on offer in the informal economy. While this furnishes the psycho-social divisions between migrant workers and Mauritanian nationals with a material base, this material base itself requires interrogation. It is, after all, no surprise that migrants, be they internal or international, are opposed to poor wages, as Deborah Potts (2010: 4) has observed. The phenomenon to be interrogated, she suggests, is the structural cause of low pay. In other words, if the informal economy sharpens colonially endowed lines of national differentiation, then this informal economy itself ought to be in the spotlight. As we will see, when we shift our focus more directly onto this informal economy – its internal social dynamics, its relationship to the state, and its global mechanisms of reproduction – the national/other divide collapses more definitively.

Drawing illegality from the racialized surplus population

At one level, the relationship of the informal economy to the production of migrant illegality in Mauritania is clear-cut. Given that the documentation required to obtain a residence permit effectively precludes those who work in the informal sector from doing so, this informal sector is itself productive of illegality. The informal sector is, moreover, the biggest employer in Mauritania, accounting for 89.4 per cent of total employment in 2019 (ILO, 2023). And while it is difficult to pin down exact numbers, it is widely accepted that the vast majority of migrants in Mauritania work informally (IOM, 2009: 16; Islamic Republic of Mauritania, 2010: 29). But what produces the informal economy in the first place? And how does this process relate to both the illegalization created by the border regime and the ethno-national and racialized distinctions detailed above?

For Kalyan Sanyal (2007), the urban informal economy is a symptom of what he calls postcolonial capitalism, which is characterized by its production of a vast sphere of 'non-capital' as a condition of capital's existence. While Marx anticipated this production of 'a population of greater extent than suffices for the average needs of the self-expansion of capital' (Marx, 1906: 419), his emphasis was upon the labour–capital relation. For Sanyal, this tension is supplemented in postcolonial contexts by that between capital and non-capital, the latter being the destination of those dispossessed and disposed through postcolonial capitalist development. Combined with the story presented in the previous chapter, the picture emerging here reflects a global state of affairs. As capital expands through forms of dispossession such as those that have occurred on the Senegal River Valley, many who fall afoul of such processes are not re-absorbed into formal wage labour. Instead, they are cast into a range of informal and semi-formal livelihood activities, such as the car washing, street vending, and domestic work by which those at the Church and elsewhere in the Nouakchott informal economy get by.

Surviving the 'urban wasteland' of postcolonial capitalism

Boubacar and I were sitting by another labour pick-up spot in Nouakchott as he was venting the above frustrations about Mauritanian Fula. This particular labour pick-up was centred around a hectic city intersection through which donkey-drawn carts, taxis, and cars in varying states of disrepair would edge their way throughout the day. The presence of a polyclinic medical dispensary just off the intersection had earned the area the name of *Clinique*. Everyday Boubacar would come to *Clinique* at eight in the morning and sit at the side of the road alongside hundreds of other bricklayers, plumbers, electricians, and other types of construction workers, in the hope of gaining a day or two's work. From the early hours of the day, the streets stemming outward from *Clinique* intersection become lined with such individuals, collectively forming a stark visual representation of the 'mass of human material ready for exploitation' by which Marx once characterized the relative surplus population (Marx, 1906: 419.) While primarily from Mali, Mauritania, and Senegal, workers here do not typically gather along national lines, unlike at SIJP in Nouadhibou.

While Boubacar found work difficult to come by at *Clinique*, this nonetheless proved more lucrative an option than remaining in his home region of Matam, which is one of the poorest and most food-insecure parts of Senegal (WFP, 2014; FAO, 2016). For this reason, Boubacar spent much of the time since his unsuccessful attempt to reach Malta circulating between Matam, where he would go during the harvest season, and Nouadhibou, where he for a time ran a small informal painting business. This rhythm of circulating between Nouadhibou and Matam was briefly interrupted in 2015, when he got a job in Senegal in a gold mine run by Randgold Resources, a gold mining company with operations across West Africa.[5] Headquartered in the Channel Islands, Randgold exemplifies a system that sees sums of money far in excess of annual inflows of development aid siphoned out of Africa each year (Ainembabazi, 2022). As far as Boubacar is concerned, such arrangements do not even translate into permanent jobs at home; his contract was terminated after a year, leading him to come back to Mauritania to restart his painting business, this time trying his luck in *Clinique*.

Varying between exploitation and expendability, Boubacar's condition is one found across the Global South. Embodied in his individual trajectory is a general grey area in which precarious and exploitative formal employment, informal labour, and structural unemployment overlap (Breman, 2016; Smith, 2016: 123; Buckley, 2020). This condition of being surplus relative to capital's reproductive needs has long been viewed in terms of its industrial reserve army function, whereby those beyond the margins of the formal labour market furnish the capitalist class with a means of depressing wages and conditions. Excluded from many of the legal protections associated with the institution of citizenship, migrant workers have long been considered an industrial reserve army par excellence (e.g. Castles and

5. Named Barrick Gold as of 2018.

Kosack, 1972). In this light, interceptions and deportations of the sort experienced by Boubacar, Khalifa, Pierre, and so many others are a violent means of artificially depressing wages within multiple sectors across southern Europe, most notably agriculture (Cross, 2013; Pradella and Cillo, 2021). And within Mauritania itself, the informal sector contributes substantially to both employment and GDP (Charmes, 1992), owing in no small part to its low wages and poor working conditions.

While this reserve army function is thus clearly at play, the informal economy is not a single homogeneous bloc, defined exclusively by its function vis-à-vis the formal labour market. As illustrated in the national antagonisms which seep into relations at the Church, this site is riven with its own inequalities, petty hierarchies, and, indeed, reserve armies and labour aristocracies (Breman, 2020). A Mauritanian national I met here named Moulaye, for instance, once expressed his frustration about how informal partnerships of the kind forged between Khalifa and Pierre have for him ended in betrayal. He told of having trained plenty of youths here, 'kids on the street who had nothing to eat, who spent days with just biscuits for dinner', only for them to ultimately stab him in the back by taking customers from him. Such grievances were commonplace among those here. And they should come as no surprise, given the paucity of formal collective bargaining rights and legal protections that characterize the informal sector. Far from being a monolithic reserve army, then, the surplus populations in the informal economy are themselves segmented and hierarchized, all the more so given the absence of any legal recognition of their labour activities (Davis, 2006: 181).

To complicate this picture further, the sectors in which day labourers such as Boubacar aspire to be hired are of course defined by the informality of the labour supply, but the clients and property owners who hire them are equally implicated in the reproduction of informality. It is for these reasons that Jan Breman (2020: 30) has challenged the scholarly habit of restricting discussion of informality to the domain of labour, drawing attention instead to the myriad means by which capital obscures and informalizes its own operations. This is certainly the case in Nouakchott, whose urban property sector has been described by Armelle Choplin (2006) as one of 'rationalised informality', in that it displays urban development patterns that are uncoded and often at odds with the statute-book, but which remain rational and structured from the point of view of the capital accumulation imperative.

If the line between formality and informality is thus far from clear-cut, the one separating the owners of capital from bearers of labour power is rather more distinct. And it is typically the latter who are illegalized as a result of the externalization process. While the 'informal sector' facilitates this illegalization, the preceding discussion will have shown that 'informal sector' is itself something of a misnomer. Undergirding this 'informal sector' in Mauritania, as elsewhere, are the dynamics of postcolonial capitalism which produce what Sanyal (2007: 45) refers to as the 'wasteland' of non-capital, inhabited by those dispossessed and marginalized through capital's reproduction.

At times, this superfluity vis-à-vis the formal labour market serves a wage depression function. At others, it appears to exceed such a function (Li, 2017: 1251; Bhattacharya, Kesar and Mehra, 2023: 149). In either case, however, when combined with the dynamics of racialized exclusion detailed in the previous section, this surplus condition generates ample human material for the border regime to illegalize. Moreover, as a result of this structural position, those within this racialized surplus population in Nouakchott are subject to bordering and dispersal operations which, in both form and function, bear a striking resemblance to the border violence that has marked so many trajectories here. In other words, the externalization process is but one aspect of a more general process of spatially managing the racialized outcasts of contemporary capitalism.

Dispersing the racialized surplus population

One day I arrived in *Clinique* to find that Boubacar had swapped his paintbrush and roller for a small selection of perfumes, which he had laid out on a tarpaulin on the curb to display to passers-by. Having found no more than three days' painting work over the space of a month, he explained, he opted to switch to street vending. 'There's no work. You spend three days, four days, weeks, without working. That's why I started this', he said, gesturing to his modest perfume display. While exemplary of the 'creative but highly precarious and insufficient ways' by which Tania Murray Li (2017: 1249) characterizes the survival strategies of those rendered surplus to capital, this decision came with its risks. As an unemployed day labourer, Boubacar largely went unnoticed by security forces during my time in Nouakchott. But his decision to enter the world of informal street vending immediately placed him in the cross-heirs of the police, who would regularly sweep through *Clinique*, confiscating the materials and goods of local street vendors. While the detentions and deportations of the *séjour* affair had 'died down' at this particular moment in the city, the antagonism between the state and migrant labour clearly lived on.

Such raids and clear-outs of Nouakchott's informal economy were a regular occurrence throughout my stay in the city. But they took on a unique intensity from February 2018 when, in the words of a *Jeune Afrique*[6] headline from the time, Nouakchott began to 'spruce up for the African Union summit in July'. Amongst the projects involved in this makeover were a new purpose-built conference centre on the northern outskirts of the city and 'a hotel and 120 villas' to house summit participants. But these were just the most conspicuous manifestations of a broader infrastructural overhaul that had been initiated in Nouakchott in anticipation of the AU summit. In the opening months of 2018, a construction drive aimed at revamping the urban landscape was rolled out. Old, cracked roads were uprooted and laid anew with fresh tarmacadam and markings. Where before, all that distinguished pavement from road were varying degrees of gravel and dust,

6. A pan-continental current affairs magazine.

now concrete had been poured and clearly delineated footpaths laid down. These pristine paths were then illuminated with freshly erected lampposts. This was a markedly uneven spatial process, but in those portions of the city earmarked for redevelopment, urban modernization had been kicked into overdrive.

As a consequence, life for Zein and his co-workers became increasingly difficult as the summit approached. The illustrious guests who would soon grace the hotel next to their car-washing site meant the aesthetics of its surrounding locale were of high priority within the urban revamp. As one worker excitedly told me, 'there are even presidents who are going to be staying there!' For this reason, municipality officials and the police were increasingly hostile to their presence as the summit approached. As far back as February, Zein and the others observed from their site as the old road in front of them was uprooted and laid afresh. But it was not long before the police began to inform them that they were no longer permitted to wash cars in this location. They soon began confiscating materials accordingly. In the line of fire were principally washing rags, brushes, and jerrycans used to store water, but the chairs, tea sets and prayers mats which make this site a space of interaction and leisure as well as work were also frequently confiscated.

At the Church too, the impending AU summit posed obstacles to car-washing work here. Things had been relatively calm after the 2016 raid, thanks to the spatial compromise between car washers and the municipality whereby the laneway east of the Church would serve as the principal car-washing site, leaving the square in front of the property vacant. Now, however, the situation had changed once more. The hotel was to finally reopen, and – much like the area around the luxury hotel just north – the presence of an informal car-washing operation and unemployed day labourers all of a sudden became unseemly. Municipality officials accompanied by the police would swoop down upon the car washers at the Church with increasing frequency, confiscating their materials and threatening them with arrest. 'Until this summit is over,' Khalifa said, leaning on an elbow on the roof of a car he was washing, 'everyone here is going to be very tired.' By the end of June, they were coming every day. 'They come and take everything and bring it to a compound', Khalifa explained, 'you have to go over there and pay to get your materials back. They make money that way.' For Zein and his co-workers too, confiscated materials could be only retrieved upon payment of a fine. As John Smith (2016: 117) observes, rather than being absent from the informal economy, the state is 'reduced to its core competencies: coercion and parasitism'.

By mid-June, the scope of the clearances had extended beyond car-washing materials and now applied to the car washers themselves. A young Malian from the Mopti region told of how a police commissioner arrived at the Church one day to inform everyone that from 23 June, the site was to be empty of people. 'All the poor people and the beggars, they want to hide them!', he observed astutely. But the 23rd came and went, and while the confiscation of materials continued apace, the clear-out never transformed into a campaign of arrest and deportation, as was the case in 2016. This was not, however, the case for Zein and others next to the luxury hotel. 'From tomorrow,' Zein said with dismay one day in June, 'we can't come here. If we're found here, either we go to prison or we get deported.

That's what they said to us.' As was the case with the clear-out at the Church in 2016, the context in which deportability becomes activated is a raid on workers in the informal economy. And sure enough, over the coming days I found the site empty. People continued to work by relocating to side streets and alleys, subsisting in less central and therefore less lucrative locales until the end of the summit. But even after the summit had finished, the raids and confiscations would continue here. Car washing here was now 'formally forbidden', Zein said they had been told. When police came, another Mauritanian worker told me, they would scatter in all directions, over the wall beneath the trees, and into a disused urban wasteland that lay beyond it.

The similarities between the policing of illegalized migrants in Mauritania and these clear-outs of urban space were not lost on those at the receiving end of these operations. After being taunted by co-workers about being afraid of the police following one particular clear-out, a young Malian retorted: 'I've already been to the Spanish border, with the Moroccan police and Spanish Guardia Civil!' Moreover, people use the same French word – *ramasse* – when speaking of the confiscation of their materials as is often used when discussing the migration control operations that often precede detention and deportation. This does not mean that these urban clear-outs should be viewed as an intentional form of immigration policing 'through the back door', of the kind Monica Varsanyi (2008) argues to be at play in public space and land-use ordinances at day labour sites in the United States. Nor is this a clear-cut case of border 'internalization', a term coined by Cecilia Menjívar (2014) to conceptualize the proliferation of migration control policies within national territories. But there is clearly scope for spill-over between these two domains, with border functions becoming apparent in the practices of urban renewal, as demonstrated by the threat of deportation issued to Zein and his co-workers. Crucially, however, while the threat of deportation was reserved for migrant workers, the scope of the urban clear-out operations extended to all who found themselves part of the racialized surplus population.

There is of course no institutional alignment between externalization and this urban development process, but nor is their convergence in form a mere coincidence. Their shared contempt for those in the informal economy is rooted in a common response to the question of how the racialized surplus population is to be dealt with. Caught within the net of this grouping in Nouakchott are both Mauritanian nationals and migrant workers. As we have seen, at the level of the informal economy in Nouakchott, many who were born on Mauritanian soil are deprived of any form of identification and are therefore suspected of being 'illegal' migrants. On a daily basis, they work and interact with migrant workers whose deportability is often contingent upon broader structural factors. This shared condition is that of the racialized surplus population, and it is the backdrop against which both migrants and Mauritanian nationals are subject to the raids and deportations compelled by externalization, as well as the clear-outs, confiscations, and enclosures associated with the urban development process. As Papadopoulos and Tsianos (2013: 189) observe, 'where migrants and precarious workers meet is in sharing the same urban spaces'; and in this case, in the repression to which this sharing of urban space exposes them.

Within this social grouping, any nationally informed distinction between insiders and outsiders collapses. In the ruins, there appears instead a tiered structure of managing, containing, and filtering the racialized surplus population, on the one hand, and the collective agency and subjectivity of this group on the other. As the past three chapters have illustrated, it is crucial to pay attention to this agency – not as a norm to be celebrated, but a social phenomenon to be interrogated. Doing so in this fashion has stripped away a Eurocentric veil in which Europe-bound movement in the region is so often cloaked, revealing instead how the variety of subject positions and aspirations that make up migrant life relate to their social and historical contexts. Ultimately, we have seen how these positions and aspirations are expressions of the more general structural condition of the racialized surplus population.

There is a paradox, however, in claiming to take people's collective agency seriously, without considering how *they understand their own* agency. How are we to take subjective aspirations and beliefs as socially constitutive without actually foregrounding the content of these aspirations and beliefs? This blind spot is especially glaring when these subjective beliefs pertain directly to the nature and scope of individual agency. In what follows, we address this paradox by tentatively exploring how those encountered in the preceding pages interpret their own agency. In doing so, we find some speculative ingredients of a non-Eurocentric universalism that will be explored in the conclusion chapter.

Coda

'*L'homme propose, Dieu dispose.*'[7] This was how Boubacar would often respond when I asked what his future plans were. This pithy formulation would seem to suggest that people have the freedom and capacity to make arrangements, plan for contingencies, and act to ensure their safety and security, but whether or not any of these efforts bear fruit is ultimately in the hands of God. This does not negate the validity of the effort – *l'homme* is the one who 'proposes' after all – but it does mean its success is contingent upon a much larger divine force. These evocations of the divine were ubiquitous within my exchanges at the Church and with people in the urban informal economy more broadly. When I asked a Senegalese electrician named Abdoulaye whether it was easy to get hired at the labour pick-up in *Clinique*, he responded: 'well, we're believers. We believe in God.' Barry, whose experience of being deported to Rosso was recounted in previous pages, proclaimed to me that 'only God knows what will happen', and that for this reason 'you can't become discouraged'. Another migrant worker, when discussing the raids and clear-out operations at the Church, remarked that 'the president has closed everything', before adding, 'but God is great'.

7. Roughly: 'Man proposes, God arranges.'

The divine would often take the form of an object of praise and thanks for those in the informal economy of Nouakchott. In the words of one migrant worker, 'God is great. As long as you have health, you should always be thankful to God.' And in response to a disgruntled car washer who had little luck attracting customers, Traoré, another Malian car washer, offered the following advice: 'it's God who does that, you should say *alḥumdulillah*.[8] You, you want to make money every day! But in reality, sometimes you make money and sometimes you don't. You should say *alḥumdulillah*.' Yousef, the Malian car washer we met above, voiced a similar sentiment, when he said with an air of quiet dignity: 'as long as I have no problems, I always thank God. At the moment I'm in this circuit. I wash cars. I get by. If ever God puts me in another circuit, I'll thank him and continue.' Clearly, one cannot investigate how the illegalized and dispossessed on the urban margins understand their own agency without acknowledging the presence of the divine within this collective subject position.

In one obvious sense, this presence serves to alleviate the sheer weight of the structures that first produce and then (attempt to) contain the racialized surplus population. As can be seen in the above examples, each allusion to being thankful to God, of God controlling all, or of knowledge ultimately resting with God, is nestled in the context of conversations whose underlying themes are precarity, violence, and abandonment. As Boubacar observed of the labour pick-up on another occasion, 'there's no work, but God is great'. In this sense, such statements correspond to Marx's much-quoted likening of religion to the 'opium of the people' (Marx, 1970: 1). Religious sentiment is, in this view, 'the sigh of the oppressed creature, the heart of a heartless world, and the soul of soulless conditions'. This sense of ennui resonates with many other exchanges I had which occupy a similar conversational register to that in which evocations of God's omniscience would arise. 'Well, we're here', or 'we're here at least',[9] are other common refrains when the question of how things are going was posed to unemployed day labourers and street vendors.

That said, it would be misguided to view these evocations of divine omniscience as mere platitudes, explicable exclusively by their psycho-social function. Of course, they pertain to the material conditions of their utterance, but they at the same time draw upon and contribute to a long-standing intellectual tradition alluded to in the second chapter of this book. This can be seen in a conversation that unfolded at the Church one day between a Mauritanian named Brahim and a Guinean called Ahmed. The discussion revolved around the question of who God helps and why. If God only helps those He loves, as Ahmed held, then one need not make an effort to improve one's circumstances. But if God loves all His creation, as Brahim argued, then it follows that the individual needs to first make some sort of effort in order for God's help to reach him: 'if you just spend the entire day at home, God won't help you', Brahim insisted, 'He loves all!' In Brahim's view,

8. 'Praise and thanks be to God.'

9. '*On est là*' and '*on est là quand même*', respectively.

then, creation (or, if one prefers, social reality) is animated by a single, unifying, compassionate force, which imbues all. Much like for Boubacar, agency[10] is not negated in this picture, but it cannot be viewed in isolation from this underlying force, of which it is an expression.

This inseparability between agency and structure is equally embedded in the exhortation to praise God, as expressed above by Traoré, Yousef, Adamah, and countless others. Being thankful for one's individual circumstances may be a coping mechanism in response to the severity of those circumstances, but it at the same time reflects the primordial ontological unity detailed in the second chapter. 'That which impels you to praise God,' Ibn 'Arabi (1980: 63) observes, 'is that essential element in you that binds you to a Divine Name.' The Islamic tradition holds that God has ninety-nine names, which are said to express reality in its totality. Often logical opposites (e.g. the First and the Last, the Manifest and the Hidden, that which Expands and that which Contracts), they express a dialectical view of reality as composed of a single fabric constituted by opposed yet interlocking forces. Each name is also said to have a correlate quality embodied in the individual, to which Ibn 'Arabi alludes here (cf. Al-Ghazali, 1992). Aside from being a means by which harsh material conditions can be reckoned with, being thankful to God thus gives voice to a latent subjective quality that expands beyond the bounds of the physical body and mind to connect with the origins of cosmological existence.

From a secular standpoint (as well as many orthodox religious ones), to cast questions of agency and structure in such a light is to utterly undo any balance between the two. Instead, individual subjectivity becomes once more subordinated to structures of domination above and beyond its control, which are all the more overpowering by virtue of their divine pretensions.[11] But this need not be the case. Indeed, from the standpoint of the Oneness of Being, the relationship between subject and object can often appear inverted. In discussing divine determination, for instance, Ibn 'Arabi (1980: 165) observes that 'the thing determined, in strict accordance with its essential state, itself determines the Determiner'. A similar zone of indistinction between subject and object is apparent in his observation that 'Knowledge has no effect on the object of knowledge, while what is known has an effect on knowledge, bestowing on it of itself what it is' (Ibn 'Arabi, 1980: 94). In other words, as an object of divine knowledge, the individual subject is constitutive of that knowledge, and thus of the Knowing Subject. Any utterance of praise merely reaffirms this contradictory state of mutual dependence.

Human free will and the divine will can thus blur and become indistinguishable, as Brahim seems to perceive in his assertion that God helps those who strive to help themselves. As Ralph Austin (1980: 33) observes in his introduction to Ibn 'Arabi's *the Bezels of Wisdom*, 'we mortals in our apparent state of otherness and

10. Or what used to be called free will, as David Graeber once pointed out.

11. The distinction between the secular and orthodox religious viewpoints in this regard is more that the former regrets this development, while the latter views it as welcome and necessary.

separation, while seemingly pawns of the divine Will, are, in reality and essentially, none other than He Who wills.' This observation flows from 'the undifferentiated and inalienable reality of the Oneness of Being, in which the whole dialectic of self-other is fused into the unimaginable and inexpressible experience of the Reality' (Austin, 1980: 33). Within the vernacular means by which those on the urban margins make sense of their reality, then, there is a latent universalism, one in which the tension between individual agency and structures of control is lucidly articulated, and ultimately dissolved in the Oneness of Being. In the conclusion chapter, we explore how this incipient universalism relates to the global juncture in which externalization unfolds.

Chapter 8

CONCLUSION

If the era of externalization in Mauritania was born of a 'crisis' of arrivals on the Canary Islands in 2006, recent years have laid bare the futility of this method of response in achieving its formal objectives. While the Atlantic Route saw a relative lull during much of the period covered in this book's empirical chapters, the number of arrivals during the early years of the Covid-19 pandemic exceeded those of 2006, with upward of 40,000 sea arrivals recorded in Spain in 2020, a precedent that was exceeded the subsequent year (UNHCR, 2021). Having been temporarily suspended in light of the Covid-19 pandemic, deportations were reinitiated in November 2020, bringing the 2003 Spain-Mauritania readmission agreement back into force (CRIDEM, 2020). The defensive machinery of externalization has also been activated in response to these arrivals, with a Guardia Civil maritime patrol aircraft being deployed to Nouakchott to provide naval and surveillance support to the Mauritanian coast guard in their interceptions of Europe-bound migrants (Ndar Info, 2020). Meanwhile, the Atlantic Route to the Canary Islands became the most lethal crossing to Europe, with the death rate overtaking that on Central and East Mediterranean passages in 2021 (Alarm Phone, 2021).

The Mauritanian government has largely abided by these imperatives of externalization, cracking down on suspected 'irregular migrants' on its territory in response to the arrivals on the Canaries (Migration-control.info, 2023). At the same time, there has been an EU push to sign Frontex status agreements with Mauritania and Senegal, which would authorize the agency to deploy its staff directly on these countries' territories, for the first time exercising its mandate on a territory not directly bordering the EU (Strik, 2023). Mauritania also, however, initiated a regularization campaign in 2022, which gave foreign nationals a temporary reprieve from the *affaire séjour* that so beleaguered many since the introduction of the biometric residence permit in 2012 (Mint Bamba, 2022). Unlike the initial rolling out of the residence permit that was detailed in Chapter 4, the fee waiver this time around was accompanied by a delivery of permits to those who enrolled, a move lauded by EU partners (Migration-control.info, 2023). All of this suggests Mauritania is becoming yet more integrated within the EU's system of externalized migration governance, more smoothly so than many other states in the region.

Indeed, Mauritania has so far been relatively insulated from a swathe of social mobilizations, military coups, and geopolitical alliance shifts that have collectively subverted European hegemony across the Sahel since 2020 (Niang, 2022). While primarily targeting the French presence in the region, these developments have also challenged the externalization process. As noted previously, the Nigerien military junta in November 2023 abrogated the anti-smuggling law 2015-036. Originally passed with extensive EU support as a measure to prevent Europe-bound movement, it had deleterious effects on the local transport industry and upon migrants whose routes were dispersed farther into the Sahara (Brachet, 2018; Border Forensics, 2023). Another prominent victim of these developments has been the G5 Sahel. A security darling of the EU throughout its existence (Lopez-Lucia, 2020), it has been in a state of crisis following Mali's withdrawal from the regional security framework in May 2022. In October 2023, Niger and Burkina Faso followed suit, shortly after declaring a parallel regional security alliance that presented itself as free from the Western interests to which the G5 Sahel was claimed to be beholden. Within these upheavals, Mauritania increasingly stands out as one of the few remaining loyal, if ambivalent, partners of Western interests in the Sahel (Ould Moctar, 2023).

These geopolitical alliance shifts and mobilizations are symptomatic of a more general waning in exclusive European dominance vis-à-vis its outside. For Achille Mbembé (2021), this juncture is one of 'late Eurocentrism', an era characterized by Europe's withdrawal from the world. In this sense, the drive to outsource migration control responsibilities to states such as Mauritania is an embodiment of European decline, of its withdrawal from the world. When Samir Amin (1988) wrote *Eurocentrism,* the Eurocentric fallacy which holds Europeanization and progress to be synonymous still held water. Today, however, it is becoming ever more untenable in the face of the lethal consequences of external migration and border policy in the Atlantic and in the Mediterranean. As Amin argues, this Eurocentric project endowed the Mediterranean with the task of naturalizing and reproducing the violent inequities of the world capitalist system, forging a metaphysical distinction between Europe and its outside. Today, on the other hand, the Mediterranean centre-periphery boundary is a site of unprecedented death and suffering, an expression of Europe's inability to live up to its own self-image. If events in the Sahel indicate a material collapse of the long-standing European reign of unhindered dominance in the region, the lethal consequences of externalization across the Mediterranean suggest a corresponding normative collapse of the claims to civilizational superiority by which EU external relations have long been conducted. In empirically detailing how externalization interacts with the social relations and structures that it encounters as it unfolds in Mauritania, this book also conceptualized this encounter's relationship to the historic juncture of waning Eurocentrism.

The externalized border in time and space

In the second part of the book, we saw that EU border externalization policies are underpinned by a dynamic tension. On the one hand, the colonial shift in territorial organization detailed in Chapter 3 fundamentally conditions the form that externalization takes in the present. At the same time, however, this colonial conditioning of externalization sits within an arc of European supremacy whose decline is manifest in the contemporary EU drive to outsource migration control responsibility. While navigating Mauritania's history through its precolonial, colonial, and postcolonial phases in this chapter, we saw that the colonial period generated a qualitative shift within the three dimensions of territorial organization that oriented its analysis – namely borders, mobility, and belonging. Prior to the turn of the nineteenth century, territories were separated by zones that could be vast in expanse, and which were liable to shift in accordance with pastoralist transhumance patterns. The colonial encounter exported the linear border form of territorial organization from the French metropole to the Senegal River Valley, before gradually spreading it throughout the territories of the Sahara. Of course, this new colonial territorial formation in the Sahara was imbued with the spatial make-up of the Emirates that inhabited this space in the precolonial era (cf. Lefevbre, 2015), but the conversion of the Senegal River into a line dividing the colonial territories of Senegal and Mauritania differed markedly from its precolonial spatial role. Moreover, this soon gave rise to a new bureaucratic approach to managing the human mobility that had long traversed the river, but which now represented migration from one colonial territory into another. When colonial efforts to regulate this mobility failed, an alternative strategy of cleaving a racial distinction between 'Moors' and 'Africans' within the territory of Mauritania was adopted, resulting in a racialization of territorial belonging that has subtly conditioned the criteria of inclusion and exclusion within the Mauritanian state ever since.

The ongoing relevance of this colonial shift in territorial organization was detailed in Chapter 4 through an examination of how the externalization process has related to these same three territorial dimensions. The externalization era commenced in Mauritania in 2006, following the spike in arrivals on the Canary Islands that year; the response initially involved a suite of securitized defensive measures, including the deployment of maritime and aerial surveillance missions and the opening of the infamous *Guantanamito* detention centre in Nouadhibou. These were soon accompanied by subtler and more deep-seated preventative measures, which embedded the concern with 'irregular migration' born in 2006 more deeply within the Mauritanian state. Embodied in the 2010 national migration strategy, these preventative measures aspired towards a 'comprehensive' approach to migration management. In forging links with broader development initiatives, however, this approach resonates with a more general 'liberal way of development', whose chief logic is the containment of the immiserated and dispossessed within the peripheries of the Global South (Duffield, 2010). While externalization's language of migration management was a new departure in Mauritania, its unfolding was conditioned by the colonial shift in territorial organization that was detailed in Chapter 3. This came to light through a return to the three dimensions of territorial organization that oriented that chapter's analysis. In the context of externalization, they took the form of an EU-supported border infrastructure project, the production of migrant illegality as a subjective experience in Mauritania, and a biometric renewal of the colonial racialization of territorial belonging. Despite exhibiting their own contemporary idiosyncrasies, the very presence of each of these processes affirms the weight of the colonial shift in territorial organization in conditioning the form that externalization takes.

None of this negates the capacity for agency on the part of Mauritanian state actors. In Chapter 3, we saw how the colonial shift in territorial organization was upheld in the postcolonial period. Be it on the northern Western Sahara border or the Senegal River Valley, Mauritanian state actors consistently reproduced the colonial assumption of a congruence between ethno-racial group and territorial borders, an assumption whose violent exclusions were epitomized in 'the events' of 1989. In other words, while at origin colonial imports, the border form of territorial distinction and the racialization of territorial belonging were each upheld by postcolonial state actors. Crucially, however, it would not be until the externalization process that the bureaucratic regulation of mobility introduced in the colonial era would acquire renewed salience. Once it did, however, Mauritanian state agency was no less apparent. The importance of the EU-Mauritania Fisheries Partnership Agreement to the EU, for instance, offered insight into leverage exercised by the Mauritanian state in the negotiation of migration control cooperation, mirroring forms of migration diplomacy and postcolonial state agency familiar from other North African contexts (El Qadim, 2014; Tsourapas, 2017). And as we analysed the externalization process through the lens of the colonial shift in territorial organization, the locus of agency shifted from EU actors to the Mauritanian state, with the latter taking increasing ownership over digital technologies associated with externalization, even earning the ire of EU officials in the process. In each instance, however, this agency reinforces rather than challenges the structure of its enactment. Externalization and its colonial conditioning are thus a mutual endeavour, albeit one animated by the opportunity structures generated by the asymmetric centre-periphery dynamics of externalization.

At the same time, however, this colonial conditioning of externalization sits within an arc of European supremacy whose decline is expressed in the contemporary EU drive to outsource migration control responsibility. This can be seen in externalization's relationship to the narrative of Eurocentrism, and in particular to its geographic imaginary of the Mediterranean centre-periphery boundary. As a precursor to the direct colonization of Mauritania at the turn of the nineteenth century, the Mediterranean centre-periphery boundary was an expression of European ascendance and African peripheralization (Rodney, 1972; Amin and Girvan, 1973). At present, however, this geographic imaginary of the Mediterranean appears to be contracting as externalization proceeds. The EU has long embodied a false, Eurocentric universalism of the kind highlighted by Samir Amin, but it took an expansive and embracing form in the 1990s, as can be seen in the liberal-peacebuilding agendas of the Barcelona Process and the Union for the Mediterranean (Youngs, 2002; Cardwell, 2011). The increasing drive towards externalization, however, is indicative of a turning inward, of a contraction in this geographic imaginary. This reconfiguration was already observable in Mauritania's relocation from the West Mediterranean to the operational zone of the Atlantic in 2006, but it accelerated following the launch of the EU Trust Fund in 2015. After this point, the Mediterranean centre-periphery boundary became a space of abandonment and death, and ceded symbolic ground to the more nakedly interest-driven geographic imaginary of the Sahel, in which Mauritania has been situated by the Trust Fund.[1]

Driving much of this is a policy desire, echoed by EU delegation officials in Nouakchott, to keep the far-right at the electoral margins in Europe. This impulse has heightened a fusion between fascist fantasy, externalization policies, and authoritarian politics beyond the EU external border (cf. Ferrer-Gallardo and Gabrielli, 2022; Ehrenreich, 2023). While the externalization process serves as temporary bulwark against Europe's history of fascism reappearing within the EU, it simultaneously upholds the colonial racial and territorial legacies that act as this history's external corollary, to take Aime Césaire description of the relationship between fascism and colonialism. In doing so, however, it heralds a waning of this colonial dominance, as the shifting geographies and politics of the Mediterranean and the Sahel suggest.

This historicization of EU external border policies offered a genealogy of the relations and structures in which those illegalized by these policies in Mauritania are embedded, but it did little to include their perspective or experience. As we saw in subsequent chapters, however, this illegalization cannot be separated from these relations and structures. Indeed, the condition of migrant illegality is ultimately drawn from the pre-existing one of the racialized surplus population. This came to light through a progressive discussion of how the illegalization of migration sits within the social relations and structures of three distinct urban locales in Mauritania.

The first of these was the northern port city of Nouadhibou, the primary departure point on Mauritanian state territory for people seeking to reach Europe. Here, externalization's recent illegalization of migration seeps into the pre-existing political economy of the city, reshaping its social relations, migratory profile, and economic development trajectory in the process. This we saw by first turning attention to the interplay between Europe-bound migrants and the policies and actors charged with controlling their movements. As we saw in the first section, this dynamic takes distinct forms, three of which were explored through the lens of the autonomy of migration, which foregrounds the agency and subjective aspirations of migrants in the analysis of migration and border regimes. The first of these was a prototypical antagonism between a Europe-bound migrant and the Guardia Civil officer, respectively embodied in Ba and Luis. A more subtle and less antagonistic interplay then emerged in the form of the professional migrant and the various embodiments of the humanitarian border in Nouadhibou. We then turned to Camara, a migrant worker who is set on reaching Europe through legal channels, but is nonetheless illegalized in Nouadhibou at the hands of local police

1. For more on how the EU's liberal peacebuilding agenda is giving way to a security approach in the Sahel, see Stambøl (2019).

forces acting in line with the diktats of externalization. Within the broad interplay between Europe-bound movement and policies aimed at stemming it, then, one finds distinct subjective orientations vis-à-vis the externalization process, with distinct elements of this process, in turn, being tasked with managing these subjective aspirations.

Nuances notwithstanding, this interplay unfolds within a broader historical trajectory and structure of relations, one that Ba's family history brought to the surface. In this light, the general juncture of European decline in which externalization unfolds comes similarly to the fore. Indeed, if we currently inhabit a world in which the global dominance of capital is increasingly decoupled from the primacy of Europe and 'the West' (Mezzadra and Neilson, 2013: 59), then Nouadhibou exemplifies this juncture. For in the early years of Mauritanian independence, Nouadhibou's industrial development was harnessed to the interests of European capital, which saw in the northern port city an export node for iron ore extracted by the European-owned MIFERMA mining operation in Zouérat (Amin, 1973; Bonté, 1975). Following the nationalization of MIFERMA and a subsequent era of structural adjustment, informalization, and neoliberalization, Nouadhibou has seen a diversification of export destinations, trade relations, and urban spatial configurations. While European capital no longer holds exclusive dominance in the city, the EU border regime facilitates these transnational capital flows through its illegalization of migrant labour, and the effective subsidization of profits this entails.

This does not mean, however, that this outcome is set in stone across the diverse sectors in which illegalized migrant workers are employed in Nouadhibou. On the contrary, Malata's refusal to be intimated into a state of timid docility demonstrated that the subjective positions woven through externalization can splinter and disrupt social relations otherwise tipped in capital's favour. Crucially, however, Europe as a destination point was firmly within the realm of Malata's prior experience. This may have emboldened his stance vis-à-vis his employer, but it certainly demonstrates the analytical need to break with the Eurocentric imaginary of the border regime, which assumes 'irregular movement' to be innately Europe-bound. This assumption animates certain strands of critical migration scholarship, and it served the chapter's analytical task of discerning externalization's relationship to Nouadhibou's postcolonial development trajectory. But as we saw, it creates barriers in the analysis of deeper colonial resonances in the present, exemplified in this case in the history of racialized territorial belonging discussed in Chapters 3.

For this reason, Chapter 6 moved to the town of Rosso on the Senegal River Valley, where attention was turned to migrants who were not on the move to Europe at the time of our encounter in the town. As the site of the colonial shift in territorial organization and the violent legacies of exclusion and expulsion that emerged from it, Rosso illustrates how the externalization process interacts with the colonial history detailed in Chapters 3 and 4. As we saw, this encounter occurs at the levels of the border itself and of migrant workers in Rosso. Regarding the former, the Rosso border has been revamped in recent years as a result of the border infrastructure project that was funded by the national migration strategy discussed in Chapter 4. In practice, however, this formal infrastructure of externalization has to reckon with two glaring social facts in Rosso. The first is the informal networks of extortion and surplus extraction associated with the town's structural peripherality, and the second is the fact that the vast majority of border crossings are by individuals who live on one side of the Senegal River and work on the other. Each of these socio-spatial facts transforms externalization as it is implemented in Rosso, the former through its conversion of international migration and border management protocol into a site of informal surplus extraction, and the latter through the legal compromise that the *laissez-passer* document forces upon the international border in Rosso.

The migrants illegalized through externalization are no less conditioned by the socio-spatial context of Rosso, and in particular its history of racialized displacement and dispossession. In marked contrast to Nouadhibou, the condition of illegality here has little bearing on subjective experience. The town is indeed a key point through which Europe-bound travellers move, and by which they are in turn deported, but there is no trace of illegalization and 'deportability' from the standpoint of situated social relations in Rosso. In order to detect what comes in its place, we took a detour from the interplay between the border and the migrant to examine the conditions of those who had been displaced during 'the events', and have since returned within the framework of a tripartite agreement between the UNHCR, Senegal, and Mauritania. Their grievances regarding exclusions from the biometric civil registry brought into relief the social tensions that continue to ensue from the colonial racialization of territorial belonging, tensions into which externalization today infuses illegality. Their demands for land restitution, meanwhile, draw attention to the process of accumulation by dispossession that accompanied 'the events', which has generated meagre, precarious, and violent employment prospects for Ali Bakar, Soro, and countless other migrant workers in the rice industry in Rosso today. From the standpoint of the situated socio-spatial relations of Rosso, then, subjective positions elsewhere marked by the border regime appear more constituted by this regional history. The final section 'returned the gaze' on externalization from the perspective of the heterogenous collective subjectivity that has emerged from the region's historically endowed social relations.

This further sharpened the distance between the externalization process and its critical analysis, as it moves firmly away from the focus on Europe-bound trajectories that animates the externalization process. At the same time, however, it risks losing sight of externalization altogether. At an empirical level, the absence of illegality in Rosso created an analytical gap that renders the analysis of externalization from the vantage point of illegalized living labour difficult. And conceptually, 'returning the gaze' from the perspective of this particular regional history highlighted the contemporary reverberations of the colonial shift in territorial organization, but it risks obscuring the global processes in which these reverberations unfold today.

For this reason, the final chapter in Nouakchott adopted a more global frame of reference, one anchored in the universalist thrust of Samir Amin. As illustrated, there are a range of experiences in Nouakchott of violent interception and deportation spanning the spatial and temporal breadth of the EU border regime in the Mediterranean and the Sahara. While this state of illegalization is also upheld by Mauritanian security forces in Nouakchott, its lived salience fluctuates in accordance with broader structural factors. This gave rise to an inquiry into who is liable to become illegalized in the first place. From such a standpoint, the reverberations of the colonial racialization of territorial belonging once again came to the fore, as Afro-Mauritanian citizens often find themselves suspected of being 'illegal' migrants and threatened with deportation. While this represents a breakdown of the national dividing line, this rupture only occurs at the level of state implementation. From the perspective of those in the informal economy, on the other hand, the national/other divide often proves quite resilient. The final section accordingly shifted the focus more directly to this informal economy, examining its relationship to postcolonial capitalism's production of vast surplus populations as a condition of its existence. Combined with the legacy of the racialization of territorial belonging, postcolonial capitalism generates a racialized surplus population from which the border regime draws the condition of migrant illegality.

As far as the logic of externalization is concerned, all those who fall within this social group need to be contained beyond the EU external border, irrespective of their subjective intentions or actual trajectory. When so framed, the externalization process appears a particularly violent, sophisticated, and large-scale aspect of a deeper process of spatially managing this racialized surplus population. As we saw in the urban development process in Nouakchott, the same people who are illegalized by the border regime are subject to bordering and dispersal tactics levelled by the state against urban informal workers, be they migrants or Mauritanian nationals. This similarity between externalization and the policing of the urban informal economy is no coincidence; it is rooted in the fact that the condition of illegality is drawn from the pre-existing condition of the racialized surplus population.

None of this would have come to light had we not taken the subjective aspirations and experience of those at the receiving end of externalization seriously. Whether it be Malata's refusal in Nouadhibou, Soro's past experience which led him to Rosso, or Boubacar's decision to change employment activities in Nouakchott, the agency of those encountered over the course of the book's ethnographic chapters has been crucial in bringing to light the underlying conditions that facilitate their illegalization. That said, any emphasis on agency rings somewhat hollow, absent a consideration of how people understand their own agency. As we saw above at the end of the previous chapter, once this question is posed, we are immediately faced with the presence of the divine, which is a constitutive dimension of how those encountered over the course of this book understand their own agency. Not only is this presence analytically important in its own right, it also has a vital epistemic bearing upon the juncture of waning Eurocentrism, of which externalization is an expression.

Towards a non-Eurocentric universalism

If externalization is a reflection of the decline in European dominance and conquest, what comes in the wake of Eurocentrism? And what might the perspective of those at the receiving end of Europe's efforts to prolong its colonial dominance over West Africa tell us in this regard? To answer these questions, let's consider more closely the theoretical underpinnings of this book's ethnographic chapters. The fifth chapter retained the Eurocentric directional assumptions of the border regime in its analysis of externalization in Nouadhibou; the sixth chapter reversed these assumptions in its exploration of Rosso, relying instead upon a 'return of the gaze' informed by the discipline of postcolonial studies; and the seventh chapter sought to transcend the limitations of each of these orientations through a global frame of reference. In this respect, each of the three is loosely informed by the following passage from *Eurocentrism*:

> The impasse is total from this point on. The contemporary world reacts to the challenge by a desperate evasion, in a two-fold culturalist involution, Eurocentric and provincialist in the West, and "inverted Eurocentric" in the Third World. More than ever, the need for a universalism capable of meeting the challenge makes a critical examination of both of these modes of thought necessary.
>
> (Amin, 1988: 78)

In adopting Europe-bound directional assumptions, the fifth chapter remained 'Eurocentric and provincialist', while the sixth chapter offered an 'inverted Eurocentric' perspective. The seventh chapter, on the other hand, sought to advance the form of non-Eurocentric universalism Amin calls for here. Crucially, however, had the final 'coda' section of this chapter been absent, it would have remained trapped in the 'total impasse' to which he refers above. To understand how this is so, we must first consider how the juncture in which externalization unfolds sits within the history of European dominance as Amin sees it.

In his view, Eurocentrism could only take hold at a moment when the European conquest of the world became materially possible. It is only after the dual turning point of 1492 that 'Europeans become conscious of the idea that the conquest of the world by their civilization is henceforth a possible objective' (Amin, 1988: 73). As a consequence, 'capitalism as a potential world system did not exist until there existed a consciousness of its conquering power' (Amin, 1988: 73). While the material potential of capitalism's global expansion coincided with European consciousness of this potential, then, the crystallization of this consciousness in the Eurocentric ideological construct preceded capitalism's physical worldwide expansion. This is a subtler, less linear rendering of the orthodox historical materialist view that depicts ideological superstructures as expressions of a prior economic base. Here, glimpses of an emergent Eurocentric superstructure can be seen before the economic base of global capitalism has been consolidated. What, in this light, is the relationship between the two today? Are similar such emergences at play in the contemporary juncture?

While capitalism's global expansion is consolidated to an unprecedented extent, the same cannot be said of the Eurocentric narrative that accompanied its centuries-long rise to dominance. Amin originally wrote *Eurocentrism* when the world was on the cusp of the unipolar order of the 1990s, a moment in which this Eurocentric narrative was reaching its fevered zenith in the form of a proclaimed end to history itself. Today, however, we inhabit a world in which 'the global dominance of capital is more and more disentangled from a world order centered on the primacy of Europe or the West' (Mezzadra and Neilson, 2013: 59). For Achille Mbembé (2017: 1) this decentring of Europe is 'the significant event, the fundamental experience, of our era'. Of course, this is not likely to be a quick or clean break. There were, after all, centuries between the crystallization of the Eurocentric ideological construct and the worldwide expansion of capitalism. This means that Eurocentrism could equally stagger forward, zombie-like, into a future whose material conditions are increasingly out of sync with its misplaced assumptions of superiority. The longer this disconnect persists, the uglier and more violent its manifestations are likely to be, as suggested in Chapter 4.

The task of crafting a 'universalism liberated from the limits of Eurocentrism' is thus more pressing but also more feasible than it was when Amin (1988: 116) originally made this call. There are, however, elements of his framework that impede its realization. As Catherine Scott (2021: 78) has observed, his is a 'thin ontology', one in which the beliefs, social life, and subjectivity of those he deems revolutionary agents do not figure centrally. This is especially unfortunate in a context such as Mauritania, where, as Erin Pettigrew (2023) shows, the realm of the invisible – encompassing spirits and divine forces – has long been constitutive of material relations and social hierarchies. Moreover, as Robbie Shilliam (2011: 13) suggests, 'the West' has forged a modern subjecthood that is built upon the erasure of the embodied and the sacred from its idealized constitution, despite the formative role they have played within it. For this reason, unearthing the sacred in the formation of modern subjectivity is crucial in evading a reductive conflation between Western modernity and global modernity, of which both Shilliam and Amin are critical (Amin, 1988: 107; Shilliam, 2011: 15).

It was for these reasons that this book explored the universalism that inheres in Ibn 'Arabi's concept of the Oneness of Being and its intuitive presence within the collective subjectivity of the racialized surplus population in Nouakchott. What might be called an intensive and an extensive dimension of this universalism can be highlighted. The intensive dimension lies in the fact of being attuned to the primordial state of the Oneness of Being as the terrain from which distinctions, oppositions, and contradictions grow. To praise the divine is to insinuate an intrinsic and unified link between the individual subject and the origins of cosmological existence. In other words, it is to transcend the parochial boundaries of history, whose purview is typically limited to all that has followed the domestication of plants and animals, and therefore excludes the vast majority of human experience, not to mention that of geological time and evolutionary biology.[2] The extensive dimension of this universalism is rooted in the fact that on a world historical scale, within what Boaventure De Sousa Santos (2007) once called a heterogeneous 'ecology of knowledges', the suffusion of the divine within everyday life is the norm, and mechanistic atheism the deviation. It is this extensive dimension to which Dipesh Chakrabarty (2001: 10) alludes in his observation that 'one empirically knows of no society in which humans have existed without gods and spirits accompanying them'.

At stake in this collective subjectivity, then, is categorically not 'universalism, conceived of as a specificity of Western modernity' (De Sousa Santos, 2018: 6). Indeed, in Amin's (1988: 15) view, the seeds of universalist philosophy were sewn in Ancient Egypt, through the concepts of an immortal soul and immanent justice that emerged within its belief system. Universalist humanism was subsequently recuperated by Hellenistic, Christian, and Islamic civilizations over the course of their development, while also manifesting independently within Hindu and Chinese philosophy. Crucially, however, while 'universalism is not the product of the modern world … it has remained only a potential before the development of European capitalism' (Amin, 1988: 72). While the concept of universalism thus precedes 'Western' capitalist modernity, for Amin it is only through the latter's worldwide expansion that it has been objectively actualized.

Today, the form this expansion takes across vast swathes of the Global South results in untold numbers being cast out of the process of capitalist development. As a result, many fall into the ranks of the racialized surplus population. They are in this sense an expression of what Amin would describe as the objective universalism entailed in the process of capital's worldwide expansion. But at the same time, the subjective means by which many navigate this structural position point to a universalism beyond the Eurocentric version that has thus far accompanied this expansion. Put differently, the universalism that inheres in the Oneness of Being emanates from a period preceding the Mediterranean's configuration as a centre-periphery boundary of the world capitalist system, but it is also rooted in the subjective experience of many of those cast out of this system today. Their expulsion occurs, moreover, at the time of a historic disconnect between the physical dynamics of global capitalism's expansion and the narrative of European civilization superiority.

Embodied in the racialized surplus population, then, and all who occupy a similar structural position in the contemporary capitalist system, is a novel and incipient universalism. It is objectively embodied by virtue of its material relationship to global capitalism's expansion, and subjectively so through its transcendence of the framework of knowledge that has thus far accompanied (and often mystified) this expansion. Amin was keenly aware of the objective aspect, but

2. This view of history has recently been broken open by David Graeber's and David Wengrow's (2021) *The Dawn of Everything*.

combined with the impasse born of the juncture in which he was writing, his 'thin ontology' arguably prevented him from discerning its subjective counterpart.[3]

This injects a small degree of optimism into a conclusion otherwise defined by a violent colonial legacy of racialized territorial belonging, the waste of global capitalism, and a border regime that sustains and consolidates the relationship between the two. For in the collective subjectivity of those on the receiving end of this arrangement, there lies an embryonic universalist outlook that surpasses the Eurocentric provincialism that has historically overlain it. Of course, this does not mean that such a universalism is itself capable of ushering in a world beyond capitalism and racialized territorial belonging. Recall that Amin does not argue capitalism's worldwide expansion to have been *caused* by the emergence of the Eurocentric ideological construct; he merely points out that the latter preceded the former. By a similar token, the presence of a universalism that transcends the epistemic framework of the capitalist system from which it originated in no way signifies a breakdown of that system. But it does mean that as and when such breakdown occurs, there is to hand an embryonic consciousness of genuinely universalist scope, which may anticipate how the structures of capitalism and racialized territorial belonging which originally spawned the Eurocentric ideological construct may be definitively buried along with it.

3. This, notwithstanding his association of the concept of the immortal soul with universalist philosophy. Moreover, in the introduction to the second edition of *Eurocentrism*, he suggests that both atheistic and deist forms of universalist humanism hold equal potential for the task of crafting a non-Eurocentric universalist ethics (Amin, 2009: 22).

REFERENCES

Abbott, S. G. (2018) 'The EU and the Middle East: From the Euro-Mediterranean Partnership to the Union for the Mediterranean', in *Conflict and Diplomacy in the Middle East: External Actors and Regional Rivalries*. Available at: https://www.e-ir.info/2018/11/28/the-eu-and-the-middle-east-from-the-euro-mediterranean-partnership-to-the-union-for-the-mediterranean/STEFANIEGEORGAKISABBOTT, NOV282018 (Accessed: 14 September 2020).

Acloque, B. (2014) 'L'idée de frontière en milieu nomade: héritage, appropriation et implications politiques actuelles (Mauritanie et Sahara occidental)', in Villasante Cervello, M. and de Beauvais, C. (eds) *Le passé colonial et les héritages actuels en Mauritanie : État des lieux de recherches nouvelles en histoire et en anthropologie sociale*. Paris: l'Harmattan, pp. 447–72.

Adamson, F. B. and Tsourapas, G. (2019) 'Migration diplomacy in world politics', *International Studies Perspectives*, 20(2), pp. 113–28. doi: 10.1093/isp/eky015.

Ahmed Salem, Z. (2001) '"Tcheb-Tchib" et Compagnie: Lexique de la survie et figures de la reussite en Mauritanie', *Politique africaine*, 82, pp. 78–100.

Ahmed Salem, Z. (2004) 'Les marges d'un État frontière: histoire régionale, clôture nationale et enjeux locale', in *Les Trajectoires d'un État-Frontier: espace, évolution politique et transformations sociales en Mauritanie*. Dakar: Conseil pour le développement de la recherche en sciences sociales en Afrique, pp. 9–45.

Ahmed Salem, Z. (2011) 'Cadre sociopolitique de la Migration en Mauritanie: principales tendances et développements récents', *Euro-Mediterranean Consortium for Applied Research on International Migration*. Available at: http://hdl.handle.net/1814/18483.

Ainembabazi, P. (2022) *Addressing Capital Flight from Africa Is a Global Task, FACT Coalition*. Available at: https://thefactcoalition.org/addressing-capital-flight-from-africa-is-a-global-task/ (Accessed: 23 January 2024).

Ajana, B. (2012) 'Biometric citizenship', *Citizenship Studies*, 16(7), pp. 851–70. doi: 10.1080/13621025.2012.669962.

Akkerman, M. (2021) 'Border Wars: The arms dealers profiting from Europe's refugee tragedy'.

Al-Akhbar (2012a) 'Mauritanie-Enrôlement des étrangers: Un Sénégalais meurt sous la bousculade', 16 October. Available at: http://fr.alakhbar.info/5083-0-Mauritanie-Enrlement-des-etrangers-Un-Senegalais-meurt-sous-la-bousculade-Video.html.

Al-Akhbar (2012b) 'Mauritanie: L'enrôlement des étrangers maintenant gratuit et le délai repoussé'. Available at: http://fr.alakhbar.info/4792-0-Mauritanie-Lenrlement-des-etrangers-maintenant-gratuit-et-le-delai-repousse.html.

Al-Akhbar (2013) 'Mauritanie: Les subsahariens se sentent discriminés au sujet de la carte de séjour', 20 March. Available at: http://www.fr.alakhbar.info/6315-0-Mauritanie-Les-subsahariens-se-sentent-discrimines-au-sujet-de-la-carte-de-sejour.html.

Al-Ghazali (1992) *The Ninety-Nine Beautiful Names of God*. Edited by D. B. Burrell and N. Daher. Cambridge: Islamic Texts Society.

Al-Naqar, U. (1969) 'Takrur: The history of a name', *Journal of African History*, 3, pp. 365–74.

Alarm Phone (2021) *Shocking number of deaths, but also growing struggles on the ground: Western Mediterranean Regional Analysis*. Available at: https://alarmphone.org/en/2021/01/29/shocking-number-of-deaths-but-also-growing-struggles-on-the-ground/#atlantic.

Altorki, S. and El-Solh, C. F. (1988) 'Introduction', in *Arab Women in the Field: studying your own society*. Syracuse: Syracuse University Press, pp. 1–23.

AMDH (2019) 'Déclaration de l'AMDH sur les rafles et refoulements des étrangers notamment subsaharien', *Initiatives News*, 23 April. Available at: https://initiativesnews.com/declaration-de-lamdh-rafles-refoulements-etrangers-notamment-subsaharien/.

Amin, S. (1973) *Neo-Colonialism in West Africa*. New York: Monthly Review Press.

Amin, S. (1974a) 'Accumulation and Development: A Theoretical Model', *Review of African Political Economy*, 1(1), pp. 9–26.

Amin, S. (1974b) 'Introduction', in *Modern Migrations in Western Africa*, p. 567. doi: 10.2307/3002350.

Amin, S. (1988) *Eurocentrism*. New York: Monthly Review Press.

Amin, S. (2009) *Eurocentrism: Modernity, Religion, and Democracy*. 2nd edition. Delhi: Aakar Books.

Amin, S. and Girvan, C. (1973) 'Underdevelopment and Dependence in Black Africa: Their historical origins and contemporary forms', *Social and Economic Studies*, 22(1), pp. 177–96.

Amnesty International (2008) *Mauritania: 'Nobody wants to have anything to do with us'*. London: Amnesty International.

Amselle, J.-L. (2022) *L'Invention du Sahel*. Vulaines-sur-Seine: Éditions du Croquant.

Anderson, B. (2003) 'Just another job? the Commodification of Domestic Labor', in Ehrenreich, B. and Hochschild, A. R. (eds) *Global woman: nannies, maids, and sex workers in the new economy*. New York: Metropolitan Books.

Anderson, B. and Ruhs, M. (2010) 'Researching Illegality and Labour Migration', *Population, Space and Place*, 16(3), pp. 175–9. doi: 10.1002/psp.594.

Anderson, B., Sharma, N. and Cynthia, W. (2009) 'Editorial: Why No Borders?', *Refuge*, 26(2), pp. 5–18.

Anderson, S. D. (2021) 'From Algiers to Timbuktu : Multi-Local Research in Colonial History across the Saharan Divide', *History in Africa*, 49, pp. 277–99. doi: 10.1017/hia.2021.22.

Andersson, R. (2014a) *Illegality, Inc. Clandestine Migration and the Business of Bordering Europe*. Oakland: University of California Press.

Andersson, R. (2014b) 'Time and the Migrant Other: European Border Controls and the Temporal Economics of Illegality', *American Anthropologist*, 116(4), pp. 795–809. doi: 10.1111/aman.12148.

Andersson, R. (2017) 'Rescued and Caught: The Humanitarian-Security Nexus at Europe's Frontiers', in De Genova, N. (ed.) *The Borders of 'Europe'*. Durham: Duke University Press, pp. 64–94.

Andersson, R. (2019) *No Go World: How Fear Is Redrawing Our Maps and Infecting Our Politics*. Berkley: University of California Press.

Anievas, A. and Nişancıoğlu, K. (2015) *How the West Came to Rule: the Geopolitical Origins of Capitalism*. London: Pluto Press.

Antil, A. (2004) 'Découpage colonial et création des frontières: le cas des frontières sahariennes et maliennes de la Mauritanie', in Salem, Z. A. (ed.) *Les Trajectoires d'un État-Frontier: espace, évolution politique et transformations sociales en Mauritanie*.

Dakar: Conseil pour le développement de la recherche en sciences sociales en Afrique, pp. 46–65.
Antil, A. and Lesourd, C. (2012) 'Une hirondelle ne fait pas le printemps. Grammaire des mobilisations sociales et politiques et retour de la question negro-mauritanienne', *L'Annee du Maghreb*, 8, pp. 407–29.
Arhin-Sam, K., Bisong, A., Jegen, L., Mounkaila, H. and Zanker, F. (2022) 'The (in)formality of mobility in the ECOWAS region: The paradoxes of free movement', *South African Journal of International Affairs*, 29(2), pp. 187–205. doi: 10.1080/10220461.2022.2084452.
Arhin-Sam, K. and Lambert, L. (2023) *Niger: Country Briefs on EU-Migration Partnerships with Third Countries*. Berlin: Brot für die Welt and Miseror.
As-Sa'd, M. A. M. W. (1989) 'Émirats et espace émiral maure: Le cas du Trârza aux XVIIIe–XIXe siècles', *Revue du Monde Musulman et de la Méditerranée*, 54, pp. 53–82. doi: 10.3406/remmm.1989.2315.
Ataç, I., Rygiel, K. and Stierl, M. (2016) 'Introduction : The Contentious Politics of Refugee and Migrant Protest and Solidarity Movements : Remaking Citizenship from the Margins', *Citizenship Studies*, 20(5), pp. 527–44. doi: 10.1080/13621025.2016.1182681.
Aung, G. (2018) 'Postcolonial capitalism and the politics of dispossession: Political trajectories in Southern Myanmar', *European Journal of East Asian Studies*, 17(2), pp. 193–227. doi: 10.1163/15700615-01702006.
Austin, R. (1980) 'Introduction', in *The Bezels of Wisdom*. New Jersey: Paulist Press.
Ba, O. (2017) 'International Justice and the Postcolonial Condition', *Africa Today*, 63(4), pp. 45–62.
Ba, O. (2022) '"The Europeans and Americans Don't Know Africa": Of Translation, Interpretation, and Extraction', *Millennium: Journal of International Studies*. doi: 10.1177/03058298211063927.
Bacchi, C. (2009) *Analysis Policy: What's the Problem Represented to Be?* French's Forest: Pearson.
Baila, A. (2016) 'La coopération entre la Mauritanie et l'Union Européenne', *Universite de Bordeaux*.
Balibar, E. (2002) 'What is a border?', in *Politics and the Other Scene*. London: Verso, pp. 75–103.
Balibar, E. (2009) 'Europe as borderland', *Environment and Planning D: Society and Space*, 27(2), pp. 190–215. doi: 10.1068/d13008.
Banaji, J. (2007) 'Islam, the Mediterranean and the rise of capitalism', *Historical Materialism*, 15(1), pp. 47–74. doi: 10.1163/156920607X171591.
Barry, B. (1972) *Le Royaume du Waalo: Le Senegal avant la Conquete*. Paris: Karthala.
Bashier, S. (2004) *Ibn Al-'Arabi's Barzakh : The Concept of the Limit and the Relationship Between God and the World*. New York: State University of New York Press.
Baxter, T. and Wenjin, P. (2016) *China's distant water fishing industry is now the largest in West Africa, Unearthed*. Available at: https://unearthed.greenpeace.org/2016/11/24/fishing-inside-chinese-mega-industry-west-africa/.
Bennoune, M. (1978) 'The Political Economy of Mauritania: Imperialism and Class Struggle', *Review of African Political Economy*, 12 (May–Aug), pp. 31–52.
Bensaâd, A. (2008) 'L'"irrégularité" de l'immigration en Mauritanie : une appréhension nouvelle, conséquence d'enjeux migratoires externes', *Euro-Mediterranean Consortium for Applied Research on International Migration*. Available at: http://hdl.handle.net/1814/10509.

Bensaâd, A. (2011) 'Aux marges du Maghreb, des tribus mondialisées', *Méditerranée*, 116, pp. 25–34.
Berriane, M., de Haas, H. and Natter, K. (2015) 'Introduction: revisiting Moroccan migrations', *Journal of North African Studies*, 20(4), pp. 503–21. doi: 10.1080/13629387.2015.1065036.
Bhambra, G. (2007) *Rethinking Modernity: Postcolonialism and the Sociological Imagination*. New York: Palgrave Macmillan.
Bhattacharya, S., Kesar, S. and Mehra, S. (2023) 'Exclusion, Surplus Population, and the Labour Question in Postcolonial Capitalism: Future Directions in Political Economy of Development', *Review of Political Economy*, 35(1), pp. 145–73. doi: 10.1080/09538259.2022.2134650.
Bhattacharyya, G. (2018) *Rethinking Racial Capitalism: Questions of Reproduction and Survival*, הארץ. London: Rowman & Littlefield Publishers.
Bialasiewicz, L. (2012) 'Off-shoring and out-sourcing the borders of Europe: Libya and EU border work in the Mediterranean', *Geopolitics*, 17(4), pp. 843–66. doi: 10.1080/14650045.2012.660579.
Bigo, D. (2000) 'When Two Become One: Internal and External Securitisations in Europe', in Kelstrup, M. and Williams, M. (eds) *International Relations Theory and the Politics of European Integration: Power, Security and Community*. London: Routledge, pp. 171–204.
Bigo, D. and Guild, E. (2005) 'Policing at a Distance: Schengen Visa Policies', in Bigo, D. and Guild, E. (eds) *Controlling Frontiers Free Movement Into and Within Europe*. Abingdon: Routledge.
Blalack, J. (2020a) 'HIDĀYAT MAN ḤĀRĀ FĪ AMR AL-NAṢ ĀRA: The Western Sahara's Missing Witness at the International Court of Justice', *L'Ouest Saharien*, 12(2). Available at: https://www.cairn.info/revue-l-ouest-saharien-2020-2.htm.
Blalack, J. (2020b) 'The Migration of Resistance and Solidarity: 'Abd Al-Qādir Al-Jazā'irī's Promotion of Hijra', *Mashriq & Mahjar Journal of Middle East and North African Migration Studies*, 7(2). doi: 10.24847/77i2020.258.
Bojadžijev, M. and Karakayalı, S. (2010) 'Recuperating the Sideshows of Capitalism: The Autonomy of Migration Today', *e-flux*, 17. Available at: https://www.e-flux.com/journal/17/67379/recuperating-the-sideshows-of-capitalism-the-autonomy-of-migration-today/.
Bonnecase, V. and Brachet, J. (2013) 'Les « crises sahéliennes » entre perceptions locales et gestions internationales', *Politique africaine*, 130(2), p. 5. doi: 10.3917/polaf.130.0005.
Bonté, P. (2008) *L'émirat de l'Adrar mauritanien*. Paris: Karthala.
Bonté, P. (1975) 'Multinational Companies and National Development: Miferma and Mauretania', *Review of African Political Economy*, 2(2), pp. 89–109. doi: 10.1080/03056247508703247.
Border Forensics (2023) *Mission accomplished? The deadly effects of border control in Niger*. Available at: https://www.borderforensics.org/investigations/niger-investigation/.
Boswell, C. (2003) 'The "External Dimension" of EU Immigration and Asylum Policy', *International Affairs*, 79(June), pp. 619–38. doi: 10.1111/1468-2346.00326.
Boulama, H., Kebe, M. K., Mbow, A., Montero, M., Sarr, A., Sarr, D. and Tixier, M. (2017) 'Axe Rosso-Nouakchott: des mobilites en danger. Rapport d'observation à la frontière Sénégal – Mauritanie'. Available at: https://www.lacimade.org/wp-content/uploads/2018/01/Loujna_Axe_Rosso_Nouakchott.pdf.
Brachet, J. (2016) 'Policing the Desert: The IOM in Libya Beyond War and Peace', *Antipode*, 48(2), pp. 272–92. doi: 10.1111/anti.12176.

Brachet, J. (2018) 'Manufacturing Smugglers: From Irregular to Clandestine Mobility in the Sahara', *Annals of the American Academy of Political and Social Science*, 676(1), pp. 16–35. doi: 10.1177/0002716217744529.
Brambilla, C. (2014) 'Shifting Italy/Libya borderscapes at the interface of EU/Africa borderland: A "genealogical" outlook from the colonial era to post-colonial scenarios', *Acme*, 13(2), pp. 220–45.
Breman, J. (2016) *On Pauperism in Present and Past*. Oxford: Oxford University Press.
Breman, J. (2020) 'Informality: The bane of the labouring poor under globalised capitalism', *The Informal Economy Revisited: Examining the Past, Envisioning the Future*, pp. 31–7.
Brenner, L. (2001) *Controlling Knowledge Religion, Power, and Schooling in a West African Muslim Society*. Bloomington: Indiana University Press.
Brower, B. C. (2022) 'An Isthmus of Modern Thought: Islam and Psychoanalysis in North Africa and the Middle East', *Modern Intellectual History*, 19(1), pp. 297–308. doi: 10.1017/S1479244320000086.
Brown, W. (2010) *Walled States, Waning Sovereignty*. Cambridge: MIT Press.
Brown, W. (2012) 'A question of agency: Africa in international politics', *Third World Quarterly*, 33(10), pp. 1889–908. doi: 10.1080/01436597.2012.728322.
Buckley, J. (2020) 'Bringing Capital Back In: Labour Informalisation and Militancy in Vietnam Declaration for SOAS PhD thesis', pp. 1–326.
Calavita, K. (1998) 'Immigration, Law and Marginalization in a Global Economy: Notes from Spain', *Law & Society Review*, 32(3), pp. 529–66.
Calavita, K. and Honig, B. (2005) 'Law, Citizenship, and the Construction of (Some) Immigrant "Others"', *Law & Social Inquiry*, 30, pp. 401–20. doi: 10.1111/j.1747-4469.2005.tb01017.x.
Cardoso, F. H. (1977) 'The Consumption of Dependency Theory in the United States', *Latin American Research Review*, 12(3), pp. 7–24.
Cardwell, P. J. (2011) 'EuroMed, European Neighbourhood Policy and the Union for the Mediterranean: Overlapping Policy Frames in the EU's Governance of the Mediterranean', *Journal of Common Market Studies*, 49(2), pp. 219–41. doi: 10.1111/j.1468-5965.2010.02119.x.
Cardwell, P. J. (2013) 'New Modes of Governance in the External Dimension of EU Migration Policy', *International Migration*, 51(6), pp. 54–66. Available at: http://www.eu-newgov.org/database/DELIV/D09D02_Special_Issue_Governance.pdf.
Carrera, S. (2007) 'The EU Border Management Strategy FRONTEX and the Challenges of Irregular Immigration in the Canary Islands', CEPS Working Document No. 261, Brussels: CEPS.
Carrera, S., Cassarino, J. P., El Qadim, N., Lahlou, M. and den Hertog, L.(2016) 'EU-Morocco Cooperation on Readmission, Borders and Protection: A Model to Follow?', *CEPS Liberty and Security in Europe series* No. 87, Brussels: CEPS.
Casas-Cortes, M. and Cobarrubias, S. (2019) 'Genealogies of contention in concentric circles: Remote migration control and its Eurocentric geographical imaginaries', in Mitchell, K., Jones, R., and Fluri, J. (eds) *Handbook on Critical Geographies of Migration*. Cheltenham: Edward Elgar, pp. 193–205. doi: 10.4337/9781786436030.00024.
Casas-Cortes, M., Cobarrubias, S. and Pickles, J. (2014) '"Good neighbours make good fences": Seahorse operations, border externalization and extra-territoriality', *European Urban and Regional Studies*, 23(3), pp. 231–51.

Casas-Cortes, M., Cobarrubias, S. and Pickles, J. (2015) 'Riding Routes and Itinerant Borders: Autonomy of Migration and Border Externalization', *Antipode*, 47(4), pp. 894–914. doi: 10.1111/anti.12148.

Cassarino, J. (2018) 'Beyond the criminalisation of migration : A non-western perspective', *International Journal of Migration and Border Studies*, 4(4), pp. 397–411.

Castañeda, H. (2010) 'Deportation Deferred: Illegality, Visibility, and Recognition in Contemporary Germany', in De Genova, N. and Peutz, N. (eds) *The Deportation Regime: Sovereignty, Space and the Freedom of Movement*. Durham: Duke University Press, pp. 245–61.

Castles, S. and Kosack, G. (1972) 'The function of labour immigration in Western European Capitalism', *New Left Review*, 73, pp. 21–44.

Cavatorta, F. and Rivetti, P. (2014) 'EU – MENA Relations from the Barcelona Process to the Arab Uprisings : A New Research Agenda', *Journal of European Integration*, 36(6), pp. 619–25. doi: 10.1080/07036337.2014.922237.

Césaire, A. (2000) *Discourse on Colonialism*. New York: Monthly Review Press.

Chacon, J. M. (2015) 'Producing Liminal Legality', *Denver University Law Review*, 92(4), pp. 710–66.

Chakrabarty, D. (2000) *Provincializing Europe: Postcolonial Thought and Historical Difference*. Princeton: Princeton University Press.

Charmes, J. (1992) 'La Contribution Du Secteur Informel a L'Emploi Et Au Produit National En Mauritanie' (October).

Chauvin, S. and Garcés-Mascareñas, B. (2012) 'Beyond informal citizenship: The new moral economy of migrant illegality', *International Political Sociology*, 6(3), pp. 241–59. doi: 10.1111/j.1749-5687.2012.00162.x.

Cherti, M. and Collyer, M. (2015) 'Immigration and Pensée d'Etat: Moroccan migration policy changes as transformation of "geopolitical culture"', *Journal of North African Studies*, 20(4), pp. 590–604. doi: 10.1080/13629387.2015.1065043.

Chipaike, R. and Knowledge, M. H. (2018) 'The question of African agency in international relations', *Cogent Social Sciences*, 4(1), pp. 1–16. doi: 10.1080/23311886.2018.1487257.

Chitonge, H. (2017) *Capitalism in Africa: The Old and the New Lyrics, Review of African Political Economy*. Available at: http://roape.net/2017/01/12/capitalism-africa-old-new-lyrics/.

Chodkiewicz, M. (1991) 'The Diffusion of Ibn 'Arabi's Doctrine', *Journal of the Muhyiddin Ibn Arabi Society*, IX. Available at: https://ibnarabisociety.org/the-diffusion-of-ibn-arabis-doctrine-michel-chodkiewicz/.

Choplin, A. (2006) 'Le foncier urbain en Afrique: Entre informel et rationnel, l'exemple de Nouakchott (Mauritanie)', *Annales de Geographie*, 115(647), pp. 69–91. doi: 10.3917/ag.647.0069.

Choplin, A. (2010) '"Quand la mer se ferme": Du transit au post-transit migratoire en Mauritanie', *Musée de l'histoire de l'immigration*, 1286(87). doi: 10.4000/hommesmigrations.1727.

Choplin, A. and Jerome, L. (2009) 'La « Mauritanie offshore ». Extraversion économique, État et sphères dirigeantes', *Politique africaine*, 2(114), pp. 87–104.

Choplin, A. and Lombard, J. (2008) 'Migrations et recompositions spatiales en Mauritanie', *Afrique Contemporaine*, 4(228), pp. 151–70.

Choplin, A. and Lombard, J. (2013) 'Stranded in Mauritania: sub-Saharan migrants in a post-transit context', in Triulzi, S. and McKenzie, R. (eds) *Long Journeys : African migrants on the road*. Leiden: Brill, pp. 69–94.

Choplin, A. and Lombard, J. (2014) 'On West African roads: Everyday mobility and exchanges between Mauritania, Senegal and Mali', *Canadian Journal of African Studies/La Revue canadienne des études africaines*, 48(1), pp. 59–75. doi: 10.1080/00083968.2014.936696.

Ciavolella, R. (2010) *Les Peuls et l'Etat en Mauritanie: Une anthropologie des marges*. Paris: Karthala.

Ciavolella, R. (2011) 'Culture nomade a l'épreuve du capital mobilité, les Peuls Fulaabe (Mauritanie)', *Mappemonde*, 103(3), pp. 1–13.

Ciccariello-Maher, G. (2017) *Decolonizing Dialectics*. Durham: Duke University Press.

Civipol (2019) *PROJET D'APPUI AU G5S POUR LA SÉCURITÉ AU SAHEL*. Available at: https://www.civipol.fr/fr/projets/projet-dappui-au-g5s-pour-la-securite-au-sahel (Accessed: 27 April 2023).

Cobarrubias, S. (2019) 'Mapping Illegality : The i-Map and the Cartopolitics of "Migration Management" at a Distance', *Antipode*, 51(3), pp. 770–94. doi: 10.1111/anti.12512.

Cobarrubias, S., Cuttitta, P., Casas-Cortés, M., Lemberg-Pedersen, M., El Qadim, N., İşleyen, B., Fine, S., Giusa, C. and Heller, C. (2023) 'Interventions on the concept of externalisation in migration and border studies', *Political Geography*, 105(3). doi: 10.1016/j.polgeo.2023.102911.

Conlon, D. (2011) 'Waiting: feminist perspectives on the spacings/timings of migrant (im)mobility', *Gender, Place and Culture*, 18(3), pp. 353–60. doi: 10.1080/0966369X.2011.566320.

Coopération: Union européenne – Mauritanie (2018) 'Revue du Partenariat entre l'Union européenne (UE) et la Mauritanie 2016–2017'.

Coopération: Union européenne – Mauritanie (2020) 'Revue du Partenariat entre l'Union européenne (UE) et la Mauritanie 2019–2020'.

Coppolani, X. (1999) *Mauritanie Saharienne: Novembre 1903 à Mai 1904 : Mission d'organisation des territoires du Tagant*. Paris: l'Harmattan.

Coutin, S. B. (2005) 'Contesting criminality: Illegal immigration and the spatialization of legality', *Theoretical Criminology*, 9(1), pp. 5–33. doi: 10.1177/1362480605046658.

Cramer, C. (2008) 'From Waging War to Peace Work: Labour and Labour Markets', in Pugh, M., Cooper, N., and Turner, M. (eds) *Whose Peace? Critical Perspectives on the Political Economy of Peacebuilding*. Basingstoke: Palgrave Macmillan.

CRIDEM (2012) 'Carte de séjour obligatoire pour tous les étrangers vivant en Mauritanie', *Cridem*, 19 May. Available at: http://cridem.org/C_Info.php?article=629624.

CRIDEM (2020) *L'Espagne se prépare à expulser des migrants illégaux vers la Mauritanie*. Available at: https://cridem.org/C_Info.php?article=742564 (Accessed: 30 November 2020).

Cross, H. M. (2011) 'Rents, rights, rejections and resistance: West African migrants, the EU's migration regime and militarisation in Mauritania', *The International Journal of Human Rights*, 15(6), pp. 827–46. doi: 10.1080/13642987.2011.586859.

Cross, H. M. (2013) *Migrants, Borders and Global Capitalism: West African Labour Mobility and EU Borders*. New York: Routledge.

Cross, H. M. (2021) *Migration Beyond Capitalism*. Cambridge: Polity Press.

Crousse, B. and Hesseling, G. (1994) 'Transformations foncières dans la vallée du Sénégal: enjeux politiques et ethniques', *Politique africaine*, 55, pp. 89–100.

Curtin, P. D. (1971) 'Jihad in West Africa: Early Phases and Inter-Relations in Mauritania and Senegal', *Journal of African History*, 12(1), pp. 11–24. doi: 10.1017/S0021853700000049.

Cuttitta, P. (2018) 'Delocalization, Humanitarianism, and Human Rights: The Mediterranean Border Between Exclusion and Inclusion', *Antipode*, 50(3), pp. 783–803. doi: 10.1111/anti.12337.
Czaika, M. and Haas, H. De (2014) 'The Effect of Visa Policies on International Migration Dynamics' (April), pp. 1–26.
Danewid, I. (2017) 'White innocence in the Black Mediterranean: Hospitality and the erasure of history', *Third World Quarterly*, 38(7), pp. 1674–89. doi: 10.1080/01436597.2017.1331123.
Danewid, I., Proglio, G., Pesarini, A., Hawthorne, C., Raeymakers, T., Saucier, K. P., Grechi, G. and Gerrand, V. (2021) 'Introduction', in Danewid, I., Proglio, G., Pesarini, A., Hawthorne, C., Raeymakers, T., Saucier, K. P., Grechi, G. and Gerrand, V. (eds) *The Black Mediterranean: Bodies, Borders and Citizenship*. London: Palgrave Macmillan, pp. 9–28.
Davis, M. (2006) *Planet of Slums*. New York: Verso.
De Chassey, F. (1978) *Mauritanie 1900–1975*. Paris: editions anthropos.
De Genova, N. (2002) 'Migrant "Illegality" and Deportability in Everyday Life', *Annual Review of Anthropology*, 31(2002), pp. 419–47. doi: 10.1146/annurev.anthro.31.040402.085432.
De Genova, N. (2004) 'The Legal Production of Mexican/Migrant " Illegality"', *Latino Studies*, 2, pp. 160–85.
De Genova, N. (2013a) 'Spectacles of migrant "illegality": The scene of exclusion, the obscene of inclusion', *Ethnic and Racial Studies*, 36(7), pp. 1180–98. doi: 10.1080/01419870.2013.783710.
De Genova, N. (2013b) '"We are of the connections": Migration, methodological nationalism, and "militant research"', *Postcolonial Studies*, 16(3), pp. 250–8. doi: 10.1080/13688790.2013.850043.
De Genova, N. (2016) 'The "Crisis" of the European Border Regime: Towards a Marxist Theory of Borders', *International Socialism*, 150, pp. 31–54.
De Genova, N. (2017a) 'Introduction: The Borders of "Europe" and the European Question', in De Genova, N. (ed.) *The Borders of 'Europe': Autonomy of Migration, Tactics of Bordering*. Durham: Duke University Press, pp. 1–36.
De Genova, N. (2017b) *The Borders of 'Europe'*. London: Duke University Press.
De Haas, H. (2008) 'The Myth of Invasion: The inconvenient Realities of African migration to Europe', *Third World Quarterly*, 29(7), pp. 1305–22. doi: 10.1080/01436590802386435.
De Haas, H. (2011) 'Irregular Migration from West Africa to the Maghreb and the European Union: An Overview of Recent Trends', *Journal of Ethnic and Migration Studies*, 37(32), pp. 522–3. doi: 10.1080/1369183X.2010.512208.
De Sousa Santos, B. (2007) 'Beyond Abyssal Thinking', *Eurozine*, June. Available at: https://www.eurozine.com/beyond-abyssal-thinking/.
De Sousa Santos, B. (2018) *The End of the Cognitive Empire: The Coming of Age of Epistemologies of the South*. Durham: Duke University Press.
Derrider, M. and Pelckmans, L. (2020) 'Reversing the Gaze: West Africa performing the EU migration-development-security nexus', *Anthropologie et développement*, 51, pp. 9–32.
Désiré-Vuillemin, G. (1997) *Histoire de la Mauritanie, dés origines à l'indépendance*. Paris: Karthala.
Di Bartolomeo, A., Fakhoury, T. and Perrin, D. (2010) *CARIM – Migration Profile: Mauritania, CARIM– Consortium for Applied Research on International Migration*.

Dia, A. (2015) 'TPMN : Le génocide biométrique continue de plus belle'. Available at: http://cridem.org/C_Info.php?article=672615.

Directorate-General for External Policies (2017) 'State of play of EU-Mauritania relations', *European Parliament*.

Directorate-General for Neighbourhood and Enlargement Negotiations (2020) 'EU delivers support to border management in Libya'. Available at: https://neighbourhood-enlargement.ec.europa.eu/news/eu-delivers-support-border-management-libya-2020-07-16_en.

Drury, M. (2019) 'On the border in northern Mauritania', *L'Annee du Maghreb*, 21. Available at: https://journals.openedition.org/anneemaghreb/5910.

Duffield, M. (2005) 'Getting savages to fight barbarians : Development, security and the colonial present', *Conflict, Security and Development*, 5(2). doi: 10.1080/14678800500170068.

Duffield, M. (2010) 'The liberal way of development and the development-security impasse: Exploring the global life-chance divide', *Security Dialogue*, 41(1), pp. 53–76. doi: 10.1177/0967010609357042.

Duffield, M. (2012) 'Challenging environments: Danger, resilience and the aid industry', *Security Dialogue*, 43(5), pp. 475–92. doi: 10.1177/0967010612457975.

Duffield, M. and Hewitt, V. (2009) 'Introduction', in Duffield, M. and Hewitt, V. (eds) *Empire, Development, & Colonialism: The Past in the Present*. Woodbridge: James Currey, pp. 1–16.

Düvell, F. (2012) 'Transit migration: A blurred and politicised concept', *Population, Space and Place*, 18, pp. 415–27. doi: 10.1002/psp.631.

Düvell, F., Triandafyllidou, A. and Vollmer, B. (2010) 'Ethical Issues in Irregular Migration Research in Europe', *Population, Space and Place*, 16(3), pp. 227–39. doi: 10.1002/psp.590.

Ehrenreich, B. (2023) 'In Melilla', *London Review of Books*. Available at: https://www.lrb.co.uk/the-paper/v45/n08/ben-ehrenreich/short-cuts.

El Qadim, N. (2014) 'Postcolonial Challenges to Migration Control: French-Moroccan Cooperation Practices on Forced Returns', *Security Dialogue*, 45(3), pp. 242–61. doi: 10.1177/0967010614533139.

El Qadim, N. and Isleyen, B. (2023) 'Border and Im/mobility Entanglements in the Mediterranean: Introduction to the Special Issue', *Environment and Planning D: Society and Space*, 41(3), pp. 3–13. doi: 10.1177/02637758231157264.

Enloe, C. (1996) 'Margins, silences and bottom rungs', *International Relations Theory: Positivism and Beyond*, pp. 186–202.

Euractiv (2022) *EU's Takuba anti-terror force quits junta-controlled Mali*. Available at: https://www.euractiv.com/section/global-europe/news/eus-takuba-anti-terror-force-quits-junta-controlled-mali/ (Accessed: 19 January 2023).

European Commission (2005) *Communication from the Commission to the Council on the Opening of Consultations with Mauritania under Article 96 of the Cotonou Agreement*. Brussels. doi: 10.1017/CBO9781107415324.004.

European Commission (2006) 'Proposal for a Council Decision concerning the conclusion of consultations with the Islamic Republic of Mauritania under Article 96 of the revised Cotonou Agreement'.

European Commission (2021) *2021: Annual Report: EU Emergency Trust Fund for Africa*.

European Commission (2022) *Five plus Five (5+5) Dialogue on Migration in the Western Mediterranean, Migration and Home Affairs, European Commission*. Available at: https://home-affairs.ec.europa.eu/networks/european-migration-network-emn/

emn-asylum-and-migration-glossary/glossary/five-plus-five-55-dialogue-migration-western-mediterranean_en (Accessed: 31 March 2023).

European External Action Service (2022) 'Holistic Strategic Review of EUTM Mali and EUCAP Sahel Mali 2022' (May).

European Parliament (2009) 'European Parliament resolution of 4 September 2008 on the coup in Mauritania', *Official Journal of the European Union* (September 2008), pp. 2008–10.

European Union (2016) 'The European Emergency Trust Fund for Stability and Addressing Root Causes of Irregular Migration and Displaced Persons in Africa: Strategic Orientation Document'. Available at: http://search.ebscohost.com/login.aspx?direct=true&db=buh&AN=16791008&site=ehost-live.

FAO (2016) 'Renforcer la résilience à Matam à travers l'éducation, les soins de santé et la productivité', *Renforcer la Résilience: les politiques a suivre*, No 7. Rome: FAO.

Fassin, D. (2011) 'Policing Borders, Producing Boundaries: The Governmentality of Immigration in Dark Times', *Annual Review of Anthropology*, 40, pp. 213–26. doi: 10.1146/annurev-anthro-081309-145847.

Feldman, D. B. (2020) 'Beyond the Border Spectacle: Global Capital, Migrant Labor, and the Specter of Liminal Legality', *Critical Sociology*, 46(4–5), pp. 729–43. doi: 10.1177/0896920519884999.

Feldman, G. (2011a) 'If ethnography is more than participant- observation, then relations are more than connections : The case for nonlocal ethnography in a world of apparatuses', *Anthropological Theory*, 11(4), pp. 375–95. doi: 10.1177/1463499611429904.

Feldman, G. (2011b) *The Migration Apparatus: Security, Labor and Policymaking in the European Union*. Redwood City: Stanford University Press.

Ferrer-Gallardo, X. (2008) 'The Spanish-Moroccan border complex: Processes of geopolitical, functional and symbolic rebordering', *Political Geography*, 27(3), pp. 301–21. doi: 10.1016/j.polgeo.2007.12.004.

Ferrer-Gallardo, X. and Gabrielli, L. (2022) 'The Ceuta Border Peripeteia: Tasting the Externalities of EU Border Externalization', *Journal of Borderlands Studies*, 37(3), pp. 645–55. doi: 10.1080/08865655.2022.2048680.

Fitzgerald, D. S. (2019) 'Remote control of migration: theorising territoriality, shared coercion, and deterrence', *Journal of Ethnic and Migration Studies*, 46(1), pp. 4–22. doi: 10.1080/1369183X.2020.1680115.

Foucault, M. (2009) *Security, Territory, Population: Lectures at the College de France, 1977–78*. New York: Palgrave Macmillan.

Frank, A. G. (1969) *Capitalism and underdevelopment in Latin America: historical studies of Chile and Brazil*. New York: Monthly Review Press.

Freedman, J. (2012) 'Analysing the gendered insecurities of migration: A case study of female Sub-Saharan African migrants in Morocco', *International Feminist Journal of Politics*, 14(1), pp. 36–55. doi: 10.1080/14616742.2011.631281.

Fresia, M. (2009) *Les Mauritaniens réfugiés au Sénégal : Une anthropologie critique de l'asile et de l'aide humanitaire*. Paris: l'Harmattan.

Frontex (2010) *Beyond the Frontiers: Frontex, the First Five Years*. Warsaw: Frontex Information and Transparency Team.

Frowd, P. M. (2014) 'The field of border control in Mauritania', *Security Dialogue*, 45, pp. 226–41. doi: 10.1177/0967010614525001.

Frowd, P. M. (2018) *Security at the Borders: Transnational Practices and Technologies in West Africa*. Cambridge: Cambridge University Press.

Frowd, P. M. (2022) 'Borderwork Creep in West Africa's Sahel', *Geopolitics*, 27(5), pp. 1331–51. doi: 10.1080/14650045.2021.1901082.
Gabrielli, L. (2016) 'Multilevel inter-regional governance of mobility between Africa and Europe: Towards a deeper and broader externalisation', *GRITIM-UPF Working Paper Series*, No. 30 (Winter). Barcelona: Grup de Recerca Interdisciplinari en Immigració.
Gaibazzi, P. (2015) *Bush Bound: Young Men and Rural Permanence in West Africa*. New York: Berghan Books.
Gaibazzi, P. (2023) 'Indirect migration management: Entangled histories of (externalized) repatriation in and beyond the Gambia's colonial legacy', *Geoforum*, Epub ahead of print, 2 March 2023. doi: 10.1016/j.geoforum.2023.103707.
Gaibazzi, P., Bellagamba, A. and Dunnwald, S. (2017) *EurAfrican Borders and Migration Management*. New York: Palgrave Macmillan.
Galvin, T. M. (2015) '"We Deport Them but They Keep Coming Back": The Normalcy of Deportation in the Daily Life of "Undocumented" Zimbabwean Migrant Workers in Botswana', *Journal of Ethnic and Migration Studies*, 41(4), pp. 617–34. doi: 10.1080/1369183X.2014.957172.
Ganga, D. and Scott, S. (2006) 'Cultural "insiders" and the issue of positionality in qualitative migration research: Moving "across" and moving "along" researcher-participant divides', *Forum: Qualitative Social Research*, 7(3). doi: 10.17169/fqs-7.3.134.
Garelli, G. and Tazzioli, M. (2013) 'Challenging the discipline of migration: Militant research in migration studies, an introduction', *Postcolonial Studies*, 16(3), pp. 245–9. doi: 10.1080/13688790.2013.850041.
Gary-Tounkara, D. (2009) 'La dispersion des Soudanais/Maliens à la fin de l'ère coloniale', *Hommes et migrations*, 1279, pp. 12–23.
Gazzotti, L. (2019) 'Deaths, Borders, and the Exception: Humanitarianism at the Spanish – Moroccan Border', *American Behavioral Scientist*, 64(4), pp. 408–35. doi: 10.1177/0002764219882990.
Gazzotti, L. (2021) *Immigration Nation: Aid, Control, and Border Politics in Morocco*. Cambridge: Cambridge University Press.
Gazzotti, L. and Hagan, M. (2021) 'Dispersal and dispossession as bordering : exploring migration governance through mobility in post-2013 Morocco', *The Journal of North African Studies*, 26(5), pp. 912–31. doi: 10.1080/13629387.2020.1800209.
Gilbert, L. and De Jong, F. (2015) 'Entanglements of Periphery and Informality in Mexico City', *International Journal of Urban and Regional Research*, 39(3), pp. 518–32. doi: 10.1111/1468-2427.12249.
Global Detention Project and ASGI (2021) 'MAURITANIA SUBMISSION TO THE UNIVERSAL PERIODIC REVIEW, 37 th SESSION OF THE UPR WORKING GROUP, January / February 2021'.
Goldschmidt, E. (2006) 'Storming the Fences: Morocco and Europe's Anti-Migration Policy', *Middle East Research and Information Project*, 239 (Summer). Available at: https://merip.org/2006/06/storming-the-fences/.
Gonzalez, C. V. (2012) 'Through the Border: Senegalese Gendered Migration to Spain', PhD thesis, University of Vancouver.
Graeber, D. and Wengrow, D. (2021) *The Dawn of Everything: A New History of Humanity*. Milton Keynes: Allen Lane.
GRDR (2014) 'Centres et Périphéries: la Moyenne Vallée du Fleuve Sénégal'.
GRDR (2018) 'Profil migratoire: Nouadhibou dans le monde, le monde a Nouadhibou'.

Greenpeace (2019) 'A waste of fish. Food security under threat from the fishmeal and fish oil industry in West Africa' (June).

Greenpeace (2021) *Feeding a Monster: How European aquaculture and animal feed industries are stealing food from West African communities.*

Griffiths, M. B. E. (2014) 'Out of Time: The Temporal Uncertainties of Refused Asylum Seekers and Immigration Detainees', *Journal of Ethnic and Migration Studies*, 40(12), pp. 1991–2009. doi: 10.1080/1369183X.2014.907737.

Gross-Wyrtzen, L. (2020) 'Contained and abandoned in the "humane" border: Black migrants' immobility and survival in Moroccan urban space', *Environment and Planning D: Society and Space*, 38(5), pp. 887–904. doi: 10.1177/0263775820922243.

Gross-Wyrtzen, L. and Gazzotti, L. (2021) 'Telling histories of the present: Postcolonial perspectives on Morocco's "radically new" migration policy', *Journal of North African Studies*, 26(5), pp. 827–43. doi: 10.1080/13629387.2020.1800204.

Grossi, T. and Lepratti, L. (2021) 'Data colonialism: The census, the map, and the software', *From the European South*, 8, pp. 47–61.

Guardia Civil (2019) *El director general visita el centro de formación del proyecto GARSI-SAHEL y el contingente de la Guardia Civil en Mauritania.* Available at: http://www.guardiacivil.es/es/prensa/noticias/6924.html (Accessed: 18 April 2019).

Guiraudon, V. (2000) 'European Integration and Migration Policy: Vertical Policy-making as Venue Shopping', *JCMS: Journal of Common Market Studies*, 38(2), pp. 251–71. doi: 10.1111/1468-5965.00219.

Gülalp, H. (1998) 'The Eurocentrism of dependency theory and the question of "authenticity": A view from Turkey', *Third World Quarterly*, 19(5), pp. 951–61. doi: 10.1080/01436599814118.

Haastrup, T., Duggan, N. and Mah, L. (2021) 'Navigating ontological (in)security in EU – Africa relations', *Global Affairs*, 7(4), pp. 541–57. doi: 10.1080/23340460.2021.1981144.

Hall, B. (2011) *A History of Race in Muslim West Africa*. Cambridge: Cambridge University Press. doi: 10.2307/j.ctv7cjw5h.5.

Hampshire, J. (2016) 'Speaking with one voice? The European Union's global approach to migration and mobility and the limits of international migration cooperation', *Journal of Ethnic and Migration Studies*, 42(4), pp. 571–86. doi: 10.1080/1369183X.2015.1103036.

Hansen, P. and Jonsson, S. (2014) *Eurafrica: The Untold History of European Integration and Colonialism*. London: Bloomsbury Publishing.

Harris, J. C. (2006) 'Raft of the Medusa', *Archives of General Psychiatry*, 63(6), pp. 602–3. doi: 10.1001/archpsyc.63.6.602.

Harvey, D. (2005) *A Brief History of Neoliberalism*. New York: Oxford University Press.

Hess, S. and Kasparek, B. (2017) 'De- and restabilising Schengen. The European border regime after the summer of migration', *Cuadernos Europeos de Deusto*, 56, pp. 47–77. doi: 10.18543/ced-56-2017pp47-77.

Heyden, S. (2022) *My Fourth Time, We Drowned*. London: Harper Collins.

Hiemstra, N. (2021) *Detain and Deport: the Chaotic US Immigration Enforcement Re.* Athens: University of Georgia Press.

Hinnant, L. (2018) 'Walk or die: Algeria strands 13,000 migrants in the Sahara', *AP News*, 25 June. Available at: https://apnews.com/9ca5592217aa4acd836b9ee091ebfc20 (Accessed: 5 August 2020).

Holden, C. (2017) 'Graduated sovereignty and global governance gaps: Special economic zones and the illicit trade in tobacco products', *Political Geography*, 59, pp. 72–81. doi: 10.1016/j.polgeo.2017.03.002.

Huysmans, J. (2000) 'The European Union and the securitization of migration', 38(5), pp. 751–77. doi: 10.1111/1468-5965.00263.
Huysmans, J. (2006) *The Politics of Insecurity: Fear, Migration and Asylum in the EU*. London: Routledge.
Ibn 'Arabi (1980) *The Bezels of Wisdom*. New Jersey: Paulist Press.
Ibrahim, M. (2005) 'The securitization of migration: A racial discourse', *International Migration*, 43(5), pp. 163–87. doi: 10.1111/j.1468-2435.2005.00345.x.
Idrissa, R. (2019) 'Dialogue in Divergence: The Impact of EU Migration Policy on West African Integration: The Cases of Nigeria, Mali, and Niger', pp. 1–48. Available at: http://www.fes.de.
Idrissa, R. (2021) 'The Sahel: A Cognitive Mapping', *New Left Review*, 132, pp. 5–39.
ILO (2023) *Country Profiles – Mauritania, ILOSTAT*. Available at: https://ilostat.ilo.org/data/country-profiles/? (Accessed: 18 January 2024).
InfoMigrants (2017) *Activists demand an end to Italy-Libya agreement*. Available at: https://www.infomigrants.net/en/post/44053/activists-demand-an-end-to-italylibya-agreement#:~:text=Italy signed a memorandum with,between Libya and the EU. (Accessed: 24 April 2023).
InfoMigrants (2023) *Migration as a weapon: Is Tunisia blackmailing the EU?* Available at: https://www.infomigrants.net/en/post/51993/migration-as-a-weapon-is-tunisia-blackmailing-the-eu (Accessed: 12 December 2023).
IOM (2009) *Migration en Mauritanie: Profil National 2009*.
IOM (2015) 'IOM Trains Mauritanian Border Police, Gendarmerie on Border Management', *IOM website*, 25 August. Available at: https://www.iom.int/news/iom-trains-mauritanian-border-police-gendarmerie-border-management.
IOM (2016) 'Cartographie et Presentation de la Gestion des Frontieres en Mauritanie'.
IOM (2018) 'Cartographie et profil socioéconomique des communautés de retour en Mauritanie (Rapport de Synthèse)', *Bureau régional pour l'Afrique de l'Ouest et du Centre de l'Organisation internationale pour les Migrations*. doi: 10.1017/CBO9781107415324.004.
IOM (2023) *5+5 Dialogue on Migration in the Western Mediterranean, IOM website*. Available at: https://www.iom.int/55-dialogue-migration-western-mediterranean.
Iskander, N. (2007) 'Informal Work and Protest : Undocumented Immigrant Activism', *British Journal of Industrial Relations*, 45(2), pp. 309–34.
Islamic Republic of Mauritania (2010a) 'Decret No 2010-150/PM portant creation, organisation et fonctionnement de l'Agence Nationale Des Titres Securises (ANRPTS)'.
Islamic Republic of Mauritania (2010b) *Stratégie nationale de gestion de la migration*.
Islamic Republic of Mauritania (2015) 'Recensement Général de la Population et de l'Habitat', *Office National de la Statistique*.
Islamic Republic of Mauritania (2017) 'Situation de l'Emploi et du Secteur Informel en Mauritanie en 2017'.
Jeune Afrique (2011) 'Recensement: ce qui met les Noirs de Mauritanie en colère', 1 October. Available at: https://www.jeuneafrique.com/153486/politique/recensement-ce-qui-met-les-noirs-de-mauritanie-en-col-re/.
Jones, R. (2011) *Violent Borders: Refugees and the Right to Move*. New York: Verso.
Jourde, C. (2004) 'Rivalites inter-(intra) ethnique et luttes symboliques dans la vallee du fleuve Senegal avant l'independance', in Salem, Z. A. (ed.) *Les Trajectoires d'un État-Frontier: espace, évolution politique et transformations sociales en Mauritanie*. Dakar: Conseil pour le développement de la recherche en sciences sociales en Afrique, pp. 66–90.

Jus, C. (2003) *Sudan francais-Mauritanie, une gépolitique coloniale (1880–1963): Tracer une ligne dans le désert*. Paris: Harmattan.
Kamara, C. S. B. and Laborie, J.-P. (1988) *L'urbanisation de la Mauritanie: Enquête dans trois villes secondaires*. Paris: La documentation française.
Kamara, C. S. B. and Leservoisier, O. (2000) 'À la quête d'une territorialité en Aftūt de Mbout sous la colonisation française (Mauritanie)', *Cahiers d'études africaine*, 40(2000), pp. 187–213.
Kane, O. (1974) 'Les Maures et le Futa-Toro au XVIIIe siècle', *Cahiers d'études africaines*, 14(54), pp. 237–52.
Karadag, S. (2019) 'Extraterritoriality of European borders to Turkey: an implementation perspective of counteractive strategies', *Comparative Migration Studies*, 7(12), pp. 1–16.
Karakayali, S. (2018) 'Volunteers: From solidarity to integration', *South Atlantic Quarterly*, 117(2), pp. 313–31. doi: 10.1215/00382876-4374856.
Katsarova, I. (2013) 'EU-Mauritania fisheries agreements', *Library of the European Parliament*.
Korvensyrjä, A. (2017) 'The Valletta Process and the Westphalian Imaginary of Migration Research', *Movements. Journal for Critical Migration and Border Regime Studies*, 3(1), pp. 191–204.
Kostakopoulou, T. (2000) 'The "protective union": Change and continuity in migration law and policy in post-Amsterdam Europe', *Journal of Common Market Studies*, 38(3), pp. 497–518. doi: 10.1111/1468-5965.00232.
Larsont, J. E. (2002) 'Informality, Illegality, and Inequality', *Yale Law & Policy Review*, 20(1), pp. 137–82.
Lavenex, S. (2006) 'Shifting up and out : The foreign policy of European immigration control Shifting Up and Out : The Foreign Policy of European Immigration Control', *West European Politics*, 29(2), pp. 329–250. doi: 10.1080/01402380500512684.
Lavenex, S. and Schimmelfennig, F. (2009) 'EU rules beyond EU borders: theorizing external governance in European politics', *Journal of European Public Policy*, 16(6), pp. 791–812.
Lawrence, B. B. (2020) 'Writing in the Eye of the Storm', *Critical Times: Interventions in Global Theory*, 3(3), pp. 456–63. doi: 10.17992/lbl.2020.04.574.
Lazarus, N. (2002) 'The fetish of "the West" in postcolonial theory', in Lazarus, N. and Bartolovich, C. (eds) *Marxism, Modernity and Postcolonial Studies*. Cambridge: Cambridge University Press, pp. 43–64.
Lecadet, C. (2017) 'Deportation Ghettoes in Mali: Expelled Migrants between State Exclusion and Self-Organization', in Gaibazzi, P., Bellagamba, A., and Dunnwald, S. (eds) *EurAfrican Borders and Migration Management*. New York: Palgrave Macmillan, pp. 109–25. doi: 10.1057/978-1-349-94972-4.
Lee, C. K. (2014) 'The Spectre of Global China', *New Left Review*, Sept(89).
Lefevbre, C. (2015) 'Introduction', in *Frontières de sable, frontières de papier: Histoire de territoires et de frontières, du jihad de Sokoto à la colonisation française du Niger, XIXe-XXe siècles*. Paris: Publications de la Sorbonne.
Lemberg-Pedersen, M. (2019) 'Manufacturing displacement. Externalization and postcoloniality in European migration control', *Global Affairs*, 5(3), pp. 247–71. doi: 10.1080/23340460.2019.1683463.
Lemberg-Pedersen, M., Stambøl, E. M., Mayblin, L., Fett, S. M. and Sahraoui, N. (2022) 'Introduction', in Lemberg-Pedersen, M., Stambøl, E. M., Mayblin, L., Fett, S. M. and Sahraoui, N. (eds) *Postcoloniality and Forced Migration*. Bristol: Bristol University Press.

Leservoisier, O. (1994) 'L'évolution foncière de la rive droite du fleuve Sénégal sous la colonisation (Mauritanie)', *Cahiers d'études africaines*, 34(133–135), pp. 55–83.
Leservoisier, O. (2012) 'Ethnicity and Interdependence: Moors and Haalpulaaren in the Senegal Valley', in McDougall, J. and Scheele, J. (eds) *Saharan Frontiers*. Indianapolis: Indiana University Press, pp. 146–61.
Leys, C. (1977) 'Underdevelopment and dependency: Critical notes', *Journal of Contemporary Asia*, 7(1), pp. 92–107. doi: 10.1080/00472337785390081.
Li, T. M. (2017) 'After Development : Surplus Population and the Politics of Entitlement', *Development and Change*, 48(June), pp. 1247–61. doi: 10.1111/dech.12344.
Lopez-Lucia, E. (2019) 'The European Union integrated and regionalised approach towards the Sahel', *Centre FrancoPaix en résolution des conflits et missions de paix* (February).
Lopez-Lucia, E. (2020) 'A tale of regional transformation: From political community to security regions the politics of security and regionalism in West Africa', *Political Geography*, 82(July), p. 102256. doi: 10.1016/j.polgeo.2020.102256.
Lutterbeck, D. (2006) 'Policing Migration in the Mediterranean', *Mediterranean Politics*, 11(1), pp. 59–82. doi: 10.1080/13629390500490411.
Maccanico, Y. (2010) 'Statewatch Analysis. EU: Controls, detention and expulsions at Europe's borders', *Statewatch Journal*, 20(3/4), pp. 1–11.
Madison, S. (2005) 'Methods: "Do I Really Need a Method" … or Deep Hanging Out?', in Madison, S. (ed.) *Critical Ethnography: Method, Ethics, and Performance*. Thousand Oakes: Sage Publications, pp. 19–50.
Maillard, M. (2018) 'Les tensions entre le Sénégal et la Mauritanie ravivées par la mort d'un pêcheur', *Le Monde*, 31 January. Available at: https://www.lemonde.fr/afrique/article/2018/01/31/les-tensions-entre-le-senegal-et-la-mauritanie-ravivees-par-la-mort-d-un-pecheur_5249891_3212.html.
Mainwaring, Ċ. (2019) *At Europe's Edge: Migration and Crisis in the Mediterranean*. Oxford: Oxford University Press.
Maisenbacher, J. (2015) 'The Political Economy of Mobility Partnerships – Structural Power in the EU's External Migration Policy', *New Political Economy*, 20(6), pp. 871–93. doi: 10.1080/13563467.2015.1041477.
Mamdani, M. (1996) *Citizen and Subject: Contemporary Africa and the Legacy of Late Colonialism*. New Jersey: Princeton University Press.
Mann, G. (2015) *From Empires to NGO's in the West African Sahel: The Road to Nongovernmentality*. Cambridge: Cambridge University Press.
Mann, G. (2021) 'French Colonialism and The Making of the Modern Sahel', in Villalon, L. (ed.) *The Oxford Handbook of the African Sahel*. Oxford: Oxford University Press, pp. 34–50. doi: 10.1093/oxfordhb/9780198816959.013.9.
Marchesin, P. (2010) *Tribus, Ethnies et Pouvoir en Mauritanie*. Paris: Karthala.
Marcus, G. E. (1995) 'Ethnography in/of the World System: The Emergence of Multi-Sited Ethnography', *Annual Review of Anthropology*, 24, pp. 95–117.
Marcussen, H. S. (1982) 'Changes in the International Division of Labour Theoretical Implications', *Acta Sociologica*, 25(1), pp. 67–78.
Marfaing, L. (2005) 'Du savoir faire sénégalais en matière de pêche sur les côtes mauritaniennes : une approche historique', *Stichproben. Wiener Zeitschrift für kritische Afrikastudien*, 8(8), pp. 69–98.
Marie, I. and Vammen, S. (2019) 'New contested borderlands: Senegalese migrants en route to Argentina', *Comparative Migration Studies*, 7(8).

Marx, K. (1870) *Marx to Sigfrid Meyer and August Vogt In New York, Marxists Internet Archive*. Available at: https://www.marxists.org/archive/marx/works/1870/letters/70_04_09.htm (Accessed: 20 July 2020).
Marx, K. (1970) 'Introduction', in O'Malley, J. (ed.) *Critique of Hegel's Philosophy of Right*. Cambridge: Cambridge University Press.
Marx, K. and Engels, F. (1906) *Capital: A Critique of Political Economy*. 3rd German. Chicago: C.H. Kerr & Company.
Mayault, I. (2017) 'Saving Senegal's sons from vanishing in European seas', *Aljazeera English*, 22 July. Available at: https://www.aljazeera.com/features/2017/07/22/saving-senegals-sons-from-vanishing-in-european-seas/?gb=true.
Mayblin, L. and Turner, J. (2021) *Migration Studies and Colonialism*. Cambridge: Polity Press.
Mbembe, A. (2000) 'At the Edge of the World: Boundaries, Territoriality, and Sovereignty in Africa', *Public Culture*, 12(1), pp. 259–84.
Mbembe, A. (2017) *Critique of Black Reason*. Durham: Duke University Press.
Mbembe, A. (2021) 'Notes on Late Eurocentrism', *Critical Inquiry*. Available at: https://critinq.wordpress.com/2021/07/01/notes-on-late-eurocentrism/.
McNevin, A. (2006) 'Political Belonging in a Neoliberal Era: The Struggle of the Sans-Papiers', *Citizenship Studies*, 10(2), pp. 135–51. doi: 10.1080/13621020600633051.
Meagher, K. (2022) 'Crisis Narratives and the African Paradox: African Informal Economies, COVID-19 and the Decolonization of Social Policy', *Development and Change*, 53(6), pp. 1200–29. doi: 10.1111/dech.12737.
Meillassoux, C. (1974) 'Development or Exploitation: Is the Sahel famine good business?', *Review of African Political Economy*, 1(1), pp. 27–33. doi: 10.1080/03056247408703235.
Menjívar, C. (2006) 'Liminal legality: Salvadoran and Guatemalan immigrants' lives in the United States', *American Journal of Sociology*, 111(4), pp. 999–1037. doi: 10.1086/499509.
Menjívar, C. (2014) 'Immigration Law Beyond Borders: Externalizing and Internalizing Border Controls in an Era of Securitization', *Annual Review of Law and Social Science*, 10(1), pp. 353–69. doi: 10.1146/annurev-lawsocsci-110413-030842.
Mezzadra, S. (2011a) 'Bringing capital back in: A materialist turn in postcolonial studies ?', *Inter-Asia Cultural Studies*, 12(1), pp. 154–64. doi: 10.1080/14649373.2011.532987.
Mezzadra, S. (2011b) 'How many histories of labour? Towards a theory of postcolonial capitalism', *Postcolonial Studies*, 14(2), pp. 151–70. doi: 10.1080/13688790.2011.563458.
Mezzadra, S. (2011c) 'The gaze of autonomy: Capitalism, migration and social struggles', in Squire, V. (ed.) *The Contested Politics of Mobility: Borderzones and Irregularity*. London: Routledge, pp. 121–42. doi: 10.4324/9780203839829.
Mezzadra, S. and Neilson, B. (2013) *Border as Method, or, the Multiplication of Labour*. Durham: Duke University Press.
Mezzadri, A. (2022) 'The Social Reproduction of Pandemic Surplus Populations and Global Development Narratives on Inequality and Informal Labour', *Development and Change*, 53(6), pp. 1230–53. doi: 10.1111/dech.12736.
Mignolo, W. D. (2002) 'The Geopolitics of Knowledge and the Colonial Difference', *South Atlantic Quarterly*, 101(1), pp. 57–96. doi: 10.2307/3720138.
Mignolo, W. D. (2012) *Local Histories/Global Designs*. Princeton: Princeton University Press.
Mignolo, W. D. and Tlostanova, M. V (2006) 'Theorizing from the Borders Shifting to Geo- and Body-Politics of Knowledge', *European Journal of Social Theory*, 9(2), pp. 205–21. doi: 10.1177/1368431006063333.

Migration-control.info (2023) *Migratory Situation in Mauritania*. Available at: https://migration-control.info/en/blog/migratory-situation-in-mauretania/ (Accessed: 20 December 2023).

Migreurop (2010) 'Prisonniers du désert: Enquête sur la situation des migrants à la frontière Mali-Mauritanie'.

Miles, W. F. S. (1994) *Hausaland Divided: Colonialism and Independence in Nigeria and Niger*. Ithaca: Cornell University Press.

Mint Bamba, O. (2022) 'Mauritanie : les ressortissants étrangers appelés à régulariser leur séjour', *Kassataya*, 24 September. Available at: https://kassataya.com/2022/09/24/mauritanie-les-ressortissants-etrangers-appeles-a-regulariser-leur-sejour/ (Accessed: 5 January 2024).

Mkandawire (2016) 'On Colonial Legacies and Social Welfare' (May). Available at: http://www.unrisd.org/80256B3C005BCCF9/(httpAuxPages)/600078C0E3281680C1257FC00032132A/$file/Mkandawire-PDRM.pdf.

Mohamedou, M. F. (2016) *Compilation et analyse des textes juridique sur la gestion des frontieres: rapport final*. Nouakchott.

Mullings, B. (1999) 'Insider or outsider, both or neither: some dilemmas of interviewing in a cross-cultural setting', *Geoforum*, 30(337–350), pp. 147–57. doi: 10.1177/1744987106056956.

N'Diaye, B. (2018) *Mauritania's Colonels: Political Leadership, Civil Military Relations, and Democratization*. London: Routledge.

Ndar Info (2020) *Lutte contre l'Immigration irrégulière : l'Espagne va mener des patrouilles aériennes depuis la Mauritanie*. Available at: https://www.ndarinfo.com/Lutte-contre-l-Immigration-irreguliere-l-Espagne-va-mener-des-patrouilles-aeriennes-depuis-la-Mauritanie_a29897.html (Accessed: 30 November 2020).

New Keywords Collective (2015) 'New Keywords : Migration and Borders', *Cultural Studies*, 29(1), pp. 55–87. doi: 10.1080/09502386.2014.891630.

Newhouse, L. (2018) 'Other Paths, Other Destinations: Towards a Manifold Reading of Mobility across Borders', *Movements. Journal for Critical Migration and Border Regime Studies*, 4(1), pp. 84–100.

Ngai, P. (2004) 'Women workers and precarious employment in Shenzhen Special Economic Zone, China', *Gender and Development*, 12(2), pp. 29–36. doi: 10.1080/13552070412331332170.

Niang, A. (2022) *Coups, insurgency and imperialism in Africa*, *Review of African Political Economy (blog)*. Available at: https://roape.net/2022/03/08/coups-insurgency-and-imperialism-in-africa/.

Nkrumah, K. (1965) *Neo-colonialism: the Last Stage of Imperialism*. London: Thoman Nelson and Sons.

Norris, H. T. (1969) 'Znaga Islam during the Seventeenth and Eighteenth Centuries', *Bulletin of the School of Oriental and African Studies, University of London*, 32(3), pp. 496–526. Available at: http://links.jstor.org/sici?sici=0041-977X(1969)32:3%3C496:ZIDTSA%3E2.0.CO;2-J.

Novak, P. (2017) 'Back to Borders', *Critical Sociology*, 43(6), pp. 847–64. doi: 10.1177/0896920516644034.

Novak, P. (2022) 'Re-producing the Humanitarian Border', *Geopolitics*, pp. 1–22. doi: 10.1080/14650045.2022.2105699.

OECD (2010) *Case no. 1: Senegal Mauritania Conflict*. Paris. Available at: https://www.oecd.org/countries/mauritania/44650413.pdf.

Ould Beidy, S. M. (2009) *L'administration territoriale et le developpement local en Mauritanie*. Paris: l'Harmattan.

Ould Cheikh, A. W. (1985) *Nomadisme, Islam et pouvoir politique dans la societe maure precoloniale [XI–XIXe]: Essai sur quelques aspects en tribalisme*. Paris: Paris V, René Descartes.

Ould Cheikh, A. W. (1995) 'Terroir, Nomadisme et Sédentarité en Mauritanie', *Academia. edu*, pp. 1–8.

Ould Cheikh, A. W. (2004) 'Espace confrerique, espace etatique: le mouridisme, le confrerisme et la frontiere mauritano-senegalaise', in Salem, Z. A. (ed.) *Les Trajectoires d'un État-Frontier: espace, évolution politique et transformations sociales en Mauritanie*. Dakar: Conseil pour le développement de la recherche en sciences sociales en Afrique, pp. 113–40.

Ould Cheikh, A. W. (2014a) 'Introduction', in *État et société en Mauritanie : Cinquante ans après l'indépendance*. Paris: Karthala, pp. 9–86.

Ould Cheikh, A. W. (2014b) 'La Mauritanie et les études postcoloniales', in Ould Cheikh, A. W. (ed.) *État et société en Mauritanie: Cinquante ans après l'Indépendance*. Paris: Karthala, pp. 121–44.

Ould Cheikh, A. W. (2018) 'Du (corps du) sujet au Sahara', pp. 497–505.

Ould Mey, M. (1996) *Global Restructuring and Peripheral States: The Carrot and the Stick in Mauritania*. Maryland: Rowman & Littlefield Publishers.

Ould Moctar, H. (2018) 'Urban Informality and the Boundaries of Belonging', *Movements*, 4(2), pp. 201–21.

Ould Moctar, H. (2020) 'The Proximity of the Past in Mauritania: EU Border Externalisation and Its Colonial Antecedents', *Anthropologie et développement*, 51, pp. 33–49.

Ould Moctar, H. (2022a) 'Autonomy within entanglements: Illegalised migrants, the EU border regime, and the political economy of Nouadhibou, Mauritania', *Environment and Planning D: Society and Space*, (December 2021), pp. 1–18. doi: 10.1177/02637758221090250.

Ould Moctar, H. (2022b) 'Saharan Sediment: Nation formation, urban development, and alter-imaginaries in the desert of Mauritania', *L'Ouest Saharien*, 17, pp. 67–90.

Ould Moctar, H. (2022c) 'The constitutive outside: EU border externalisation, regional histories, and social dynamics in the Senegal River Valley', *Geoforum* (September). doi: 10.1016/j.geoforum.2022.09.009.

Ould Moctar, H. (2023) *Biden Just Gave Mauritania a Pass on Slavery and Forced Labor, World Politics Review*. Available at: https://www.worldpoliticsreview.com/us-mauritania-slavery-sahel/ (Accessed: 13 December 2023).

Ould Saad, M. E. M. (2004) 'Le fleuve Sénégal comme enjeu régional de la "colonisation agricole" à l'après-barrage (1818-2001)', in Ould Ahmed Salem, Z. (ed.) *Les Trajectoires d'un État-Frontier: espace, évolution politique et transformations sociales en Mauritanie*. Dakar: Conseil pour le développement de la recherche en sciences sociales en Afrique, pp. 90–112.

Oviedo Hernandez, D. and Titheridge, H. (2016) 'Mobilities of the periphery: Informality, access and social exclusion in the urban fringe in Colombia', *Journal of Transport Geography*, 55, pp. 152–64. doi: 10.1016/j.jtrangeo.2015.12.004.

Pallister-Wilkins, P. (2015) 'The humanitarian politics of european border policing: Frontex and border police in evros', *International Political Sociology*, 9(1), pp. 53–69. doi: 10.1111/ips.12076.

Pallister-Wilkins, P. (2017) 'The Tensions of the Ceuta and Melilla Border Fences', in Gaibazzi, P., Bellagamba, A., and Dunnwald, S. (eds) *Eur-African Borders and*

Migration Management: Political Cultures, Contested Spaces, and Ordinary Lives. New York: Palgrave Macmillan, pp. 63–81. doi: 10.1057/978-1-349-94972-4.
Panzeri, P. A. (2010) 'Sixth Interparliamentary meeting between the Delegation for relations with the Maghreb countries and the Delegation from the Mauritanian Parliament', *European Parliament*. Available at: https://www.mendeley.com/catalogue/ff3e73ba-a74b-33b9-82e0-1d65a289c573/.
Papadopoulos, D. and Tsianos, V. (2013) 'After citizenship : autonomy of migration, organisational ontology and mobile commons', *Citizenship Studies*, 17(2), pp. 178–96. doi: 10.1080/13621025.2013.780736.
Parker, R. (1991) 'The Senegal-Mauritania Conflict of 1989: A Fragile Equilibrium', *The Journal of Modern African Studies*, 29(1), pp. 155–71. doi: 10.1017/S0022278X00020784.
Parreñas, R. (2001) *Servants of Globalization: Migration and Domestic Work*. Stanford: Stanford University Press.
Peers, S. (2006) 'Key Legislative Developments on Migration in the European Union', *European Journal of Migration and Law*, 8(3), pp. 321–56.
Peregil, F. (2015) 'How Spain halted the flow of migrants from Mauritania to the Canary Islands', *El Pais*.
Perkowski, N. (2016) 'Humanitarianism, Human Rights, and Security in EUropean Border Governance: The Case of Frontex'.
Petras, J. (1982) 'The "Peripheral State": Continuity and change in the international division of labour', *Journal of Contemporary Asia*, 12(4), pp. 415–31. doi: 10.1080/00472338285390351.
Petras, J. and Brill, H. (1985) 'The Tyranny of Globalism', *Journal of Contemporary Asia*, 15(4), pp. 403–20. doi: 10.1080/00472338580000271.
Pettigrew, E. (2007) 'Colonizing the Mahadra: Language, Identity and Power in Mauritania Under French Control', *Ufahamu: A Journal of African Studies*, 33(2–3), pp. 62–89.
Pettigrew, E. (2019) *The History of Islam in Mauritania, Oxford Research. Encyclopedia of African History*. Oxford: Oxford University Press. doi: 10.1093/acrefore/9780190277734.013.628.
Pettigrew, E. (2021) 'Du baptême du nouvel aéroport de Nouakchott à la réforme constitutionnelle (fin 2015–2018). Politiques de l'histoire en Mauritanie', *L'Année du Maghreb*. doi: 10.4000/anneemaghreb.5880.
Pettigrew, E. (2023) *Invoking the Invisible in the Sahara: Islam, Spiritual Mediation and Social Change*. Cambridge: Cambridge University Press.
Pezzani, L. and Heller, C. (2013) 'A disobedient gaze: strategic interventions in the knowledge(s) of maritime borders', *Postcolonial Studies*, 16(3), pp. 289–98. doi: 10.1080/13688790.2013.850047.
Portail Mauritanien du Developpement (2009) *Appel d'offre: Acquérir une solution biométrique pour la fiabilisation et la sécurisation des documents nationaux*. Available at: http://www.pmd.mr/pages/fr/services/tender?idTender=C0DD803C-7F00-0001-00A9-2DAF63E4794E.
Potts, D. (2010) *Circular Migration in Zimbabwe and Contemporary Sub-Saharan Africa*. New York: Boydell & Brewer.
Poutignat, P. and Streiff-fénart, J. (2008) 'Nouadhibou « ville de transit » ? Le rapport d'une ville à ses étrangers dans le contexte des politiques de contrôle des frontières de l'Europe', *Revue europeenne des migrations internationales*, 24(2), pp. 193–217.
Poutignat, P. and Streiff-Fénart, J. (2010) 'The development of a policy of migration in Mauritania: process, issues and actors', in Geiger, M. and Pécoud, A. (eds) *The Politics of International Migration Management*. Hampshire: Palgrave Macmillan, pp. 1–16.

Pradella, L. and Cillo, R. (2021) 'Bordering the surplus population across the Mediterranean: Imperialism and unfree labour in Libya and the Italian countryside', *Geoforum*, 126(June 2020), pp. 483–94. doi: 10.1016/j.geoforum.2020.06.019.
Pradella, L. and Taghdisi Rad, S. (2017) 'Libya and Europe: Imperialism, crisis and migration', *Third World Quarterly*, 38(11), pp. 2411–27. doi: 10.1080/01436597.2017.1350819.
Prebisch, R. (1959) 'Commercial Policy in the Underdeveloped Countries', *The American Economic Review*, 49(2), pp. 251–73.
Quijano, A. (2000) 'Coloniality of Power, Eurocentrism, and Latin America', *Coloniality of Power, Eurocentrism, and Latin America*, 1(3), pp. 181–224. doi: 10.1515/9780822388883-009.
Radio France Internationale (2011) 'Recensement en Mauritanie: les réfugiés rapatriés entrent dans la boucle', 14 October. Available at: http://www.rfi.fr/afrique/20111014-recensement-mauritanie-negro-mauritaniens-entrent-boucle.
Raineri, L. (2021) 'The Bioeconomy of Sahel Borders: Informal Practices of Revenue and Data Extraction', *Geopolitics*, pp. 1470–91. doi: 10.1080/14650045.2020.1868439.
Raineri, L. and Strazzari, F. (2019) '(B)ordering Hybrid Security? EU Stabilisation Practices in the Sahara-Sahel Region', *Ethnopolitics*, 18(5), pp. 544–59. doi: 10.1080/17449057.2019.1640509.
Reeves, M. (2013) 'Clean fake: Authenticating documents and persons in migrant Moscow', *American Ethnologist*, 40(3), pp. 508–24. doi: 10.1111/amet.12036.
République Islamique de Mauritanie (2013) 'Loi n°2013-001 du 02 janvier 2013 portant création de la zone franche de Nouadhibou', pp. 52–71.
Reslow, N. and Vink, M. (2015) 'Three-Level Games in EU External Migration Policy: Negotiating Mobility Partnerships in West Africa*', *Journal of Common Market Studies*, 53(4), pp. 857–74. doi: 10.1111/jcms.12233.
Robinson, D. (1975) 'The Islamic Revolution of Futa Toro', *The International Journal of African Historical Studies*, 8(2), pp. 185–221.
Robinson, D. (2000) *Paths of Accommodation: Muslim societies and French colonial authorities in Senegal and Mauritania, 1890-1920*. Paris: Karthala.
Rodney, W. (1972) *How Europe Underdeveloped Africa*. Cape Town: Pambazuka Press.
Rodney, W. (1973) 'Some Questions on Development', in *How Europe Underdeveloped Africa*, pp. 1–44.
Roman, E. (2018) 'Reversing the Perspective : How European Stakeholders React to Migration Policy Frames of Southern Mediterranean Counterparts' *MedReset Working Papers*, No 18(July). Rome: Istituto Affari Internazionali.
Rossi, B. (2015) 'Kinetocracy: The government of mobility at the desert's edge', *Mobility Makes States: Migration and Power in Africa*, pp. 147–68.
Sabaratnam, M. (2017) *Decolonising Intervention: International Statebuilding in Mozambique*. London: Rowman & Littlefield Publishers.
Samba Sylla, N. (2021) 'Fighting monetary colonialism in francophone Africa', *Review of African Political Economy*, 48(167), pp. 32–49. doi: 10.1080/03056244.2021.1878123.
Samers, M. (2004) 'An emerging geopolitics of "illegal" immigration in the European union', *European Journal of Migration and Law*, 6(1), pp. 27–45. doi: 10.1163/1571816041518750.
Sánchez, O. (2003) 'The rise and fall of the dependency movement: Does It inform underdevelopment today?', *EIAL: Estudios Interdisciplinarios de America Latina y el Caribe*, 14(2), pp. 31–50.
Santoir, C. (1990) 'Le conflit mauritano-senegalais: la genese. Le cas des Peul de la haute vallee du Senegal', *Cahiers – ORSTOM, Serie Sciences Humaines*, 26(4), pp. 553–76.

Sanyal, K. (2007) *Rethinking Capitalist Development: Primitive Accumulation, Governmentality, and Post-colonial Capitalism*. London: Routledge.
Sassen, S. (2005) 'The Repositioning of Citizenship and Alienage: Emergent Subjects and Spaces for Politics?', *Globalizations*, 2(1), pp. 79–94. Available at: papers2://publication/uuid/2E3C6DCA-BE9D-4DD9-947D-F9B2D0005260.
Savio Vammen, I. M., Cold-Ravnkilde, S. M. and Lucht, H. (2022) 'Borderwork in the Expanded EU-African Borderlands', *Geopolitics*, 27(5), pp. 1317–30. doi: 10.1080/14650045.2022.2008734.
Schain, M. a. (2009) 'The state strikes back: Immigration policy in the European union', *European Journal of International Law*, 20(1), pp. 93–109. doi: 10.1093/ejil/chp001.
Scheel, S. (2013) 'Autonomy of Migration Despite Its Securitisation ? Facing the Terms and Conditions of Biometric Rebordering', *Millennium: Journal of International Studies*, 41(3), pp. 575–600. doi: 10.1177/0305829813484186.
Scott, C. (2021) 'The gender of dependency theory: Women as workers, from neocolonialism in West Africa to the implosion of contemporary capitalism', *Review of African Political Economy*, 48(167), pp. 66–81. doi: 10.1080/03056244.2021.1882415.
Scott, J. C. (1998) *Seeing Like a State: How certain schemes to improve the human condition have failed*. London: Yale University Press.
Seddon, D. (1996) 'The political economy of Mauritania: An introduction', *Review of African Political Economy*, 23(68), pp. 1987–214. doi: 10.1080/03056249608704195.
Sen, A. (1983) '8. 1 The Sahel, the Drought, and the Famine', 9780191596(3).
Shilliam, R. (2011) *International Relations and Non-Western Thought*. Edited by R. Shilliam. New York: Routledge.
Sileye, B. (2016) 'Enquête – reportage Mauritanie : Scandale de corruption au cœur des contrôles de migrants', *Radio Mauritanie Internationale*, 29 November. Available at: http://rmi-info.com/enquete-reportage-mauritanie-scandale-de-corruption-coeur-controles-de-migrants/ (Accessed: 7 March 2018).
Smirl, L. (2015) *Spaces of Aid: How Cars, Compounds and Hotels Shape Humanitarianism*. London: Zed Books.
Smith, J. (2016) *Imperialism in the 21st Century: the Globalization of Production, Super-Exploitation and the Crisis of Capitalism*. New York: Monthly Review Press.
Soto Bermant, L. (2015) 'The Myth of Resistance: Rethinking the "Informal" Economy in a Mediterranean Border Enclave', *Journal of Borderlands Studies*, 30(2), pp. 263–78. doi: 10.1080/08865655.2015.1046993.
Squire, V. (2010) 'The Contested Politics of Mobility', *The Contested Politics of Mobility*, pp. 1–26. doi: 10.4324/9780203839829.
Squire, V. (2017) 'Governing migration through death in Europe and the US: Identification, burial and the crisis of modern humanism', *European Journal of International Relations*, 23(3), pp. 513–32. doi: 10.1177/1354066116668662.
Stambøl, E. M. (2019) 'The Rise of Crimefare Europe: Fighting Migrant Smuggling in West Africa', *European Foreign Affairs Review*, 24(3), pp. 287–307. doi: 10.54648/EERR2019026.
Stambøl, E. M. and Jegen, L. (2022) 'Colonial Continuities and the Commodification of Mobility Policing', in Lemberg-Pedersen, M., Stambøl, E. M., Mayblin, L., Fett, S. M. and Sahraoui, N. (eds) *Postcoloniality and Forced Migration*. Bristol: Bristol University Press, pp. 76–92. doi: 10.2307/j.ctv2tjdgxr.10.
Statewatch (2021) *Senegal: Biometric population database will facilitate deportations, restricted EU document confirms*. Available at: https://www.statewatch.org/news/2021/

april/senegal-biometric-population-database-will-facilitate-deportations-restricted-eu-document-confirms/ (Accessed: 28 March 2023).

Statewatch (2022) *Plan for Frontex to deploy 'vessels, surveillance equipment, and carry out operational tasks' in Senegal and Mauritania.* Available at: https://www.statewatch.org/news/2022/july/eu-tracking-the-pact-plan-for-frontex-to-deploy-vessels-surveillance-equipment-and-carry-out-operational-tasks-in-senegal-and-mauritania/.

Stewart, C. (1973) *Islam and Social Order in Mauritania: a case study from the 19th century.* Oxford: Oxford University Press.

Stierl, M. (2016) 'Contestations in death – the role of grief in migration struggles struggles', *Citizenship Studies*, 20(2), pp. 173–91. doi: 10.1080/13621025.2015.1132571.

Stierl, M. (2019) 'Of Migrant Slaves and Underground Railroads: Movement, Containment, Freedom', *American Behavioral Scientist.* doi: 10.1177/0002764219883006.

Stierl, M. (2020a) 'Do No Harm? The Impact of Policy on Migration Scholarship', *Environment and Planning C: Politics and Space.* doi: 10.1177/2399654420965567.

Stierl, M. (2020b) 'Reimagining EUrope through the Governance of Migration', *International Political Sociology*, pp. 252–69. doi: 10.1093/ips/olaa007.

Stierl, M. and Dadusc, D. (2021) 'The "Covid excuse": EUropean border violence in the Mediterranean Sea', *Ethnic and Racial Studies*, 45(8), pp. 1453–74. doi: 10.1080/01419870.2021.1977367.

Stierl, M. and Tazzioli, M. (2021) 'Toward an Archive of Migrant Struggles: Critique and the Materiality of Justice', in Grappi, G. (ed.) *Migration and the Contested Politics of Justice: Europe and the Global Dimension.* Lonon: Routledge, pp. 227–39.

Stock, I. (2012) 'Gender and the dynamics of mobility: Reflections on African migrant mothers and "transit migration" in Morocco', *Ethnic and Racial Studies*, 35(9), pp. 1577–95. doi: 10.1080/01419870.2011.594175.

Stock, I., Üstübici, A. and Schultz, S. U. (2019) 'Externalization at work : Responses to migration policies from the Global South', *Comparative Migration Studies*, 7(48), pp. 1–9.

Streiff-Fénart, J. (2011) 'The manufacture of transit: Border control, urban trends, and migrant trajectories in Nouahibou (Mauritania)', in Streiff-Fénart, J. and Segatti, A. (eds) *The challenge of the threshold: Border closures and migration movements in Africa.* Maryland: Rowman & Littlefield Publishers, pp. 39–56.

Strik, T. (2010) 'Readmission agreements: A mechanism for returning irregular migrants', *Parliamentary Assembly of the Council of Europe.*

Strik, T. (2023) *Draft report on a European Parliament recommendation concerning negotiations on a status agreement between the European Union and the Islamic Republic of Mauritania on operational activities carried out by the European Border and Coast Guard Agency (Frontex) in the Islamic Republic of Mauritania.* Brussels: Committee on Civil Liberties, Justice and Home Affairs. Series volume: 2087(INI).

Sylla, A. and Cold-Ravnkilde, S. M. (2022) 'En Route to Europe? The Anti-politics of Deportation from North Africa to Mali', *Geopolitics*, 27(5), pp. 1390–409. doi: 10.1080/14650045.2021.1995358.

Tasamba, J. (2023) 'Mauritania says 3 escaped terrorist prisoners killed', *Andalou Agency*, 12 March. Available at: https://www.aa.com.tr/en/africa/mauritania-says-3-escaped-terrorist-prisoners-killed/2843288.

Tazzioli, M. (2014) *Spaces of Governmentality: Autonomous Migration and the Arab Uprisings.* London: Rowman & Littlefield Publishers.

Tazzioli, M. and De Genova, N. (2020) 'Kidnapping migrants as a tactic of border enforcement', *EPD: Society and Space.* doi: 10.1177/0263775820925492.

Testas, A. (2001) 'Maghreb-EU Migration: Interdependence, Remittances, the Labour Market and Implications for Economic Development', *Mediterranean Politics*, 6(3), pp. 64–80. doi: 10.1080/713604532.

Thandika Mkandawire (2001) 'Thinking about Developmental States in Africa', *Cambridge Journal of Economics*, 25, pp. 289–313. Available at: https://watermark.silverchair.com/250289.pdf?token=AQECAHi208BE49Ooan9kkhW_Ercy7Dm3ZL_9Cf3qfKAc485ysgAAAlYwggJSBgkqhkiG9w0BBwagggJDMIICPwIBADCCAjgGCSqGSIb3DQEHATAeBglghkgBZQMEAS4wEQQMlyMbAfJjMfjp6zyOAgEQgIICCTjdXK7hMX8Kth35bOmyfIkMGLDGSRBKLNHSkYSr8coWamlR.

The New Arab (2017) 'Refugees in Libya held in "concentration camp-like" condition', November. Available at: https://english.alaraby.co.uk/features/refugees-libya-held-concentration-camp-conditions.

Thiam, M. (2016) 'Les migrants damnés de la terre mauritanienne : La carte de séjour ou le parcours du combattant', *Le Calame*, 18 August. Available at: http://www.lecalame.info/?q=node/4392.

Thurston, A. (2020) 'Mauritania', in *Jihadists of North Africa and the Sahel*. Cambridge: Cambridge University Press, pp. 264–310.

Torpey, J. (1998) 'Coming and Going : On the State Monopolization of the Legitimate "Means of Movement"', *Sociological Theory*, 16(3), pp. 239–59.

Traore, A. S. (2013) *Mauritanie : carte de séjour pour les étrangers, Reines d'Afrique*. Available at: http://awaseydou.mondoblog.org/2013/11/07/mauritanie-carte-de-jours-pour-les-etrangers/ (Accessed: 10 May 2019).

Triandafyllidou, A. (2014) 'Multi-levelling and externalizing migration and asylum: Lessons from the southern European islands', *Island Studies Journal*, 9(1), pp. 7–22. doi: 10.1108/17568691211248757.

Tsianos, V. and Karakayalı, S. (2010) 'Transnational Migration and the Emergence of the European Border Regime: An Ethnographic Analysis', *European Journal of Social Theory*, 13(3), pp. 373–87. doi: 10.1177/1368431010371761.

Tsing, A. L. (1994) 'From the Margins', *Cultural Anthropology*, 9(3), pp. 279–97.

Tsourapas, G. (2017) 'Migration diplomacy in the Global South: cooperation, coercion and issue linkage in Gaddafi's Libya', *Third World Quarterly*, 38(10), pp. 2367–86. doi: 10.1080/01436597.2017.1350102.

Tuqoui, J.-P. (2006) 'Nouadhibou chef-lieu de l'émigration sauvage', *Le Monde*, 22 March. Available at: https://www.lemonde.fr/a-la-une/article/2006/03/22/nouadhibou-chef-lieu-de-l-emigration-sauvage_753437_3208.html.

Tyszler, E. (2019) 'From controlling mobilities to control over women's bodies: Gendered effects of EU border externalization in Morocco', *Comparative Migration Studies*, 7(1). doi: 10.1186/s40878-019-0128-4.

UNCTAD (2016) 'Fish-rich Mauritania looks to its Nouadhibou port for a brighter future'. Available at: https://unctad.org/news/fish-rich-mauritania-looks-its-nouadhibou-port-brighter-future.

UNHCR (2009) 'Mauritania', *UNHCR Global Report*, pp. 161–5.

UNHCR (2021) *Operational Data Portal: Refugee Situations (Spain)*. Available at: https://data2.unhcr.org/en/country/esp.

United Nations Human Rights Council (2016) 'Report of the Special Rapporteur on torture and other cruel, inhuman or degrading treatment or punishment on his mission to Mauritania', *Trends in the Sciences*, 7(8), pp. 44–5. doi: 10.5363/tits.7.8_44.

UNU-Merit and Governance, Mgs. Maastricht G. S. of (2017) 'Mauritania: Migration Profile', *Study on Migration Routes in West and Central Africa* (August). Available at: http://migration.merit.unu.edu/publications/reports/ (Accessed: 4 December 2020).
Van Baar, H. (2016) 'Evictability and the Biopolitical Bordering of Europe', *Antipode*, 49(1), pp. 212–30. doi: 10.1111/anti.12260.
Varsanyi, M. W. (2008) 'Immigration Policing Through the Backdoor: City Ordinances, the "right to the City," and the Exclusion of Undocumented Day Laborers', *Urban Geography*, 29(1), pp. 29–52. doi: 10.2747/0272-3638.29.1.29.
Vigneswaran, D. and Quirk, J. (2015) *Mobility Makes States*. Philadelphia: University of Pennsylvania Press.
Villasante Cervello, M. and Taylor, R. (2014) 'Introduction', in Villasante Cervello, M. and de Beauvais, C. (eds) *Le passé colonial et les héritages actuels en Mauritanie: État des lieux de recherches nouvelles en histoire et en anthropologie sociale*. Paris: l'Harmattan, pp. 25–56.
Vium, C. (2016) 'Enacting immanent potentialities: Tcheb-tchib strategies at the centre of the urban fringe in Nouakchott, Mauritania', *Singapore Journal of Tropical Geography*, 37(3), pp. 343–62. doi: 10.1111/sjtg.12158.
Vives, L. (2011) 'White Europe: An alternative reading of the Southern EU border', *Geopolitica(s)*, 2(1), pp. 51–70.
Vives, L. (2017) 'The European Union–West African sea border: Anti-immigration strategies and territoriality', *European Urban and Regional Studies*, 24(2), pp. 209–24. doi: 10.1177/0969776416631790.
Vives, L. (2021) 'Death at Sea : Dismantling the Spanish Search and Rescue System', *Geopolitics*, 28(2), pp. 641–66. doi: 10.1080/14650045.2021.1973438.
Wade, R. (2018) 'The Developmental State: Dead or Alive?', *Development and Change*, 49(2), pp. 518–46. doi: 10.1111/dech.12381.
Wai, Z. (2018) 'Africa in/and International Relations : An Introduction', in *Recentering Africa in International Relations: Beyond Lack, Peripherality, and Failure*, pp. 1–29.
Walia, H. (2021) *Border and Rule: Global Migration, Capitalism, and the Rise of Racist Nationalism*. Chicago: Haymarket Books.
Walker, R. B. J. (2002) 'After the Future: Enclosures, Connections, Politics', in Falk, R., Ruis, L. and Walker, R. B. J. (eds) *Reframing the International*. London: Routledge.
Wallerstein, I. (1974) 'The rise and future demise of the world capitalist system: Concepts for comparative analysis', *Comparative Studies in Society and History*, 16(4), pp. 113–39. doi: 10.4324/9780429269400-14.
Walsh, A. (2017) *Far-right ship to return migrants to Africa, DW*. Available at: https://www.dw.com/en/defend-europe-identitarians-charter-a-ship-to-return-migrants-to-africa/a-39702947 (Accessed: 24 April 2023).
Walters, W. (2010) 'Foucault and Frontiers: Notes on the Birth of the Humanitarian Border', in Bröckling, U., Krasmann, S., and Lemke, T. (eds) *Governmentality: Current Issues and Future Challenges*. Routledge, pp. 138–164.
Walters, W. (2015) 'Reflections on Migration and Governmentality', *Movements. Journal for Critical Migration and Border Regime Studies*, 1(1), pp. 1–25.
Wane, S. (2016) 'Un migrant malien tué lors d'un contrôle d'identité à Nouakchott', *Radio Mauritanie Internationale*, 12 May. Available at: http://rmi-info.com/un-migrant-malien-tue-lors-dun-controle-didentite-a-nouakchott/.
Webb, J. L. A. (1995) *Desert Frontier: Ecological and Economical Change Among the Western Sahel*. Madison: University of Wisconsin Press.

WFP (2014) 'Analyse Globale de la Vulnérabilité, de la Sécurité Alimentaire et de la Nutrition (AGVSAN): Sénégal'. Available at: http://documents.wfp.org/stellent/groups/public/documents/ena/wfp266798.pdf.

Whitehouse, B. (2012) *Migrants and Strangers in an African City: Exile, Dignity and Belonging*. Bloomington: Indiana University Press.

Wildman, S. (2017) *A European alt-right group wants to take to the sea to stop rescuers from saving migrants, Vox*. Available at: https://www.vox.com/world/2017/7/6/15804196/generation-identity-identitarians-alt-right-migration-islam-refugees-europe (Accessed: 24 April 2023).

Williams, J. M. (2015) 'From humanitarian exceptionalism to contingent care: Care and enforcement at the humanitarian border', *Political Geography*, 47, pp. 11–20. doi: 10.1016/j.polgeo.2015.01.001.

Wimmer, A. and Glick Schiller, N. (2002) 'Methodological nationalism and beyond: Nation-state building, migration and the social sciences', *Global Networks*, 2(4), pp. 301–34. doi: 10.1111/1471-0374.00043.

Wolff, S. (2008) 'Border management in the Mediterranean: Internal, external and ethical challenges external and ethical challenges', *Cambridge Review of International Affairs*, 21(2), pp. 253–71. doi: 10.1080/09557570802021030.

Wolff, S. (2014) 'The Politics of Negotiating EU Readmission Agreements: Insights from Morocco and Turkey', *European Journal of Migration and Law*, 16, pp. 69–95.

Wolff, S. and Pawlak, P. (2017) 'The Southern Mediterranean: A testing ground and a litmus test for EU JHA policies and research?', in Ripoll Servent, A. and Trauner, F. (eds) *The Routledge Handbook of Justice and Home Affairs Research*. London: Routledge, pp. 310–22.

Yatera, S. (1996) *Mauritania: Immigration et developpement dans la vallee du fleuve Senegal*. Paris: l'Harmattan.

Youngs, R. (2002) 'The European Union and Democracy Promotion in the Mediterranean: A New or Disingenuous Strategy?', *Democratization*, 9(January 2015), pp. 40–62. doi: 10.1080/714000237.

Zaiotti, R. (2007) 'Of Friends and Fences: Europe's Neighbourhood Policy and the "Gated Community Syndrome"', *Journal of European Integration*, 29(2), pp. 143–62. doi: 10.1080/07036330701252581.

Zampagni, F. (2017) 'The Making of the Schengen Regime: Visa Filtering at the Italian Embassy', in Gaibazzi, P., Dunnwald, S., and Bellagamba, A. (eds) *EurAfrican Borders and Migration Management: Political Cultures, Contested Spaces, Ordinary Lives*. New York: Palgrave Macmillan, pp. 157–79.

Zaun, N. and Nantermoz, O. (2022) 'The use of pseudo-causal narratives in EU policies: The case of the European Union Emergency Trust Fund for Africa', *Journal of European Public Policy*, 29(4), pp. 510–29. doi: 10.1080/13501763.2021.1881583.

INDEX

www.ingramcontent.com/pod-product-compliance
Ingram Content Group UK Ltd.
Pitfield, Milton Keynes, MK11 3LW, UK
UKHW020655180425
457554UK00004B/82

9 781350 37678